MWF: 1-2:30

Introduction to

MODERN ALGEBRA

Third Edition

Introduction to MODERN ALGEBRA

Neal H. McCoy
Professor Emeritus of Mathematics
Smith College

Allyn and Bacon, Inc.
Boston · London · Sydney · Toronto

470 ATLANTIC AVENUE, BOSTON, MASSACHUSETTS 02210

PRINTED IN THE UNITED STATES OF AMERICA

Library of Congress Cataloging in Publication Data

McCoy, Neal Henry, 1905–
Introduction to modern algebra.

Bibliography: p.
1. Algebra, Abstract. I. Title.
QA162.M3 1975 512 74-19140

ISBN 0-205-04545-6

10 9 8 7 6 85 84 83 82 81

Dedicated to the memory of my son

Paul

Contents

Preface

This edition, in common with the first two editions, is intended as a text for a *first* course in abstract algebra. My goal has continued to be that of making the exposition as clear and simple as possible, but sufficiently precise and thorough to furnish an honest introduction to the methods and results of abstract algebra.

A considerably expanded version of the second edition has been published under the title *Fundamentals of Abstract Algebra*, Allyn and Bacon, 1972. In particular, that book contains a substantial extension of the material on linear algebra as well as additional material on groups, rings, and fields. That book may therefore be of help in meeting the requirements of those who found that the second edition did not have sufficient material to meet their needs. This present third edition goes somewhat in the other direction. It has become apparent that a good many instructors who adopted the book as a text used, in fact, only a few chapters. In particular, many of them did not use the material on linear algebra since students are increasingly taking an entirely separate course in that subject. Accordingly, all linear algebra has been omitted in this edition.

In addition to the deletion of linear algebra, the principal changes in subject matter are as follows. A brief introductory chapter on the logical structure of mathematics has been included since it has been my experience that even well-prepared students can profit from a quick review of this point of view. Chapter 8 on finite abelian groups has been added for those who may want to do a little more with groups. This chapter is taken without essential change from *Fundamentals of Abstract Algebra.* Although this material is presented as soon as the students will have had the necessary background, at the option of the instructor it may be omitted entirely or postponed until later in the course. The remaining chapters make no use of the results of this chapter. A final chapter on factorization in integral domains has been added for those who might desire a little additional abstract algebra. This topic seems particularly appropriate since it has close connections with several topics that have been covered earlier. In particular, it serves to unify some of the similar results about integers and polynomials which have been presented in previous chapters.

A number of relatively minor changes have also been incorporated. We may mention that Chapter 1 has been expanded somewhat, and parts of it have been rewritten. Some additional exercises have been introduced here and there throughout the book. A short and selective bibliography has been added, as well as notes and references at the ends of several chapters.

A few sections are marked with an asterisk, both in the table of contents and where they occur in the text. Although the material of these sections may be of interest in itself, it does not contribute to the development of the main ideas of the text and these sections may therefore be omitted at the option of the instructor.

It may be appropriate to make one additional remark about Chapter 1. Although I have found it to be desirable to introduce these preliminary topics as early as possible, no doubt some instructors might prefer to wait to present them until they are actually needed. This procedure should be possible with only a little care on the part of the instructor.

I am indebted to a considerable number of people in deciding what revisions would be desirable in this edition. In addition to a number of fairly short, but valuable, comments, rather extensive and constructive suggestions were made by Professors Thomas R. Berger, William G. Roughead, and Harvey Wolff. I am further indebted to Professors Henry Frandsen, Stephen Dragosh, Robert M. McConnel, and Harvey Wolff for reading an essentially final version of the manuscript. Each of them made fairly detailed suggestions and also caught some slips, thus helping to improve the quality of the exposition at various points. Although not all suggestions were followed, they were all seriously considered and were of great value.

Northampton, Massachusetts **Neal H. McCoy**

Introduction to
MODERN ALGEBRA

Chapter 0

The Logical Structure of Mathematics

The outstanding characteristic of modern abstract algebra, and indeed also of many other branches of modern mathematics, is its extensive use of what is known as the postulational or axiomatic method. The method itself is not new, since it was used by Euclid (about 300 B.C.) in his construction of geometry as a deductive science. However, in many ways the modern viewpoint is quite different from Euclid's, and the power of the method did not become apparent until the twentieth century.

At the present time, most students of mathematics have had some introduction to the postulational method before taking a course such as the one for which this book is designed as a text. However, in this brief preliminary chapter we present, primarily for review purposes, a brief outline of the basic ideas. Many illustrations will occur throughout this book.

Undefined Terms and Postulates

If one uses a dictionary to try to find the meaning of an unknown word, the definition is necessarily given in terms of another word (or words). If this word is also unknown, one must then try to find *its* meaning. A little thought will convince one that it is impossible for a dictionary to define all words; that is, a person must already know the meanings of some words or the dictionary is of no help.

In a mathematical system we avoid this problem by starting with a few *undefined terms* and make no attempt to define them. We then list some postulates or axioms (we shall use these words interchangeably), involving the undefined terms. The important fact is that the postulates tell us *all* that we need to know about the undefined terms.

As an example, in a study of plane geometry we might take *point*, *line*, *set*, and *between* as undefined terms.* Then among the postulates might be such as the following:

(1) A line is a set of points.
(2) There is exactly one line which contains two given distinct points.
(3) Given two distinct points on a line, there is another point on the line which is between them.

The first precise and modern treatment of plane geometry was due to Hilbert (1862–1943). In his approach, there were five undefined terms and fifteen postulates. We shall not pursue this example any further except to emphasize that it does not matter at all what one *thinks of* when the undefined terms are used—the important thing is that they have the properties given in the postulates.

Definitions

After one has the undefined terms, the postulates may be presented; in general, they describe assumed relationships between the undefined terms. Later on, it may be useful to define some additional terms. Of course, these definitions must depend only on the undefined terms and the postulates (or something which may have already been proved by use of the postulates). By way of illustration, in the above example of plane geometry we might define a new term *segment* as follows:

If A and B are distinct points on a line, the set of points consisting of the points A and B and all the points of the line between A and B is called a *segment* AB.

Note that postulates (1) and (3) assure us that there *are* some points on a segment. Observe, also, that this definition uses all the undefined terms mentioned above, namely, *point*, *line*, *set*, *between*.

Statements and Theorems

By a *statement* we shall mean a declaratory statement for which it is meaningful to say that it is either "true" or "false." We explicitly deny that a statement may be both true and false. Euclid considered the postulates to be statements which were obviously true in the physical

* The word *set* may not be used explicitly in some treatments of geometry, but the concept underlies most of mathematics and we shall use it whenever we wish. This concept will be discussed briefly in the first section of the following chapter.

world. The modern point of view is to consider the postulates of a system as statements to which the label *true* is assigned; that is, we are not directly concerned with physical reality, but the postulates are taken as true by the "rules of the game."

A statement which has been proved to be true in a given system is usually called a *theorem* (or a *proposition*). However, if the truth of a statement follows almost immediately from some theorem, the statement is usually called a *corollary* of the theorem. Likewise, a result which is primarily obtained as an aid to the proof of a later theorem, and which may otherwise not be of great interest in itself, is often called a *lemma*.

Implications

As an example of the next concept to be introduced, let ABC be a triangle in a given plane and let us consider the following statements:

(p) Triangle ABC is an equilateral triangle.
(q) Triangle ABC is an isosceles triangle.

Now consider the statement, "If p is true, then q is true." A statement of this form is called an *implication*. If it is a true implication, as in this example, we often say that *p implies q*. Many of the theorems in mathematics assert the truth of some implication. In proving a theorem of the form, "If p is true, then q is true," we call the statement p the *hypothesis* and the statement q the *conclusion*.

Suppose that p and q are given statements and that we are interested in the implication, "If p is true, then q is true." Associated with this implication in a natural way are three other implications as follows:

(Converse) If q is true, then p is true.
(Inverse) If p is false, then q is false.
(Contrapositive) If q is false, then p is false.

We might emphasize that we are not asserting anything about whether these are true implications. In the example in which p and q are the statements made above about a triangle, the implication "If p is true, then q is true" is a true implication. Neither the converse nor the inverse of this implication is true, but the contrapositive is readily seen to be true. This illustrates an important logical point as follows. It is always true that *an implication and its contrapositive are either both true or both false*. A little thought will make this convincing. As a hint, suppose that the implication "If p is true, then q is true" is a true implication. Assume, now, that q is false. Then we must have that p is false, for if p were true, the given implication would show that q is true. Since q cannot be both true and false, we conclude that p must be false. A similar agument shows that if the contrapositive of a given implication is true, so is the given implication.

If r and s are statements such that both of the implications "If r is true, then s is true" and "If s is true, then r is true" are true, it is convenient to say that the statements r and s are *equivalent*. We have observed above that an implication and its contrapositive are equivalent. Thus, in order to prove some specified implication, we may just as well prove its contrapositive.

It may be observed that the contrapositive of the contrapositive of a given implication is the given implication. Moreover, the converse and the inverse of a given implication are contrapositives of each other.

Hypothesis and Conclusion

As we have indicated, a theorem is often of the form:

Theorem A. *If p is true, then q is true.*

Here we start with the truth of p as hypothesis, and prove the conclusion that q is true. However, we could equally well prove the contrapositive equivalent theorem:

Theorem A′. *If q is false, then p is false.*

In this case, we would take as hypothesis the fact that q is false, and the conclusion would be that p is false.

It may happen in a given situation that a proof of one of these theorems seems more natural than a proof of the other. In that case, we should of course always choose the more convenient form since the two theorems are equivalent.

Indirect Proofs

Sometimes an *indirect* proof is convenient. If we should try to give an indirect proof of Theorem A above, we would take as hypothesis that p is true and that q is false. The goal would then be to show that this hypothesis leads to a contradiction of one of the postulates or of something which has already been proved from the postulates. This contradiction would show that we cannot have simultaneously p true and q false, and this establishes the desired result.

"If and Only If"

It is not unusual to find a theorem which takes the following form:

Theorem B. *Statement p is true if and only if statement q is true.*

By this "if and only if" wording is meant that both of the following implications are true:

If q is true, then p is true.
If q is false, then p is false.

It will be observed that the contrapositive form of the second of these is

If p is true, then q is true.

Thus Theorem B asserts that p and q are equivalent statements. A theorem involving the words "if and only if" therefore always requires the proof of two implications. Of course, either or both of them may be proved in the contrapositive form if desired.

Although we have here been speaking of implications, we may remark that a *definition* is always to be interpreted as an "if and only if" statement, although this particular grammatical construction is frequently not used in formulating a definition. Thus, the postulate

(1) A line is a set of points,

mentioned above, is by no means a definition of a line since it is not true that a set of points is necessarily a line. This postulate simply states *one* of the properties of a line. We have, in fact, suggested that *line* is frequently taken as one of the undefined terms in plane geometry.

This concludes our quick survey of the main features of so-called abstract mathematics. Illustrations of all of the concepts introduced above will occur frequently in later chapters of this book.

Chapter 1

Some Fundamental Concepts

In this chapter we present a few basic concepts to be used repeatedly, and introduce some convenient notation. Although the reader may very well have previously met some, or even all, of these concepts, they are so fundamental for our purposes that it seems desirable to present them here in some detail.

1.1 SETS

The concept of *set* (class, collection, aggregate) is fundamental in mathematics as it is in everyday life. A related concept is that of *element* of a set. We make no attempt to define these terms; that is, we shall consider them to be undefined terms in our system. However, we shall presently give some examples that will illustrate the sense in which they are being used.

First of all, we may say that a set is made up of elements. In order to give an example of a set we need, therefore, to exhibit its elements or to give some rule that will specify its elements. We shall often find it convenient to denote sets by capital letters and elements of sets by lower-case letters. If a is an element of the set A, we may indicate this fact by writing $a \in A$ (read, "a is an element of A"). Also, $a \notin A$ will mean that a is not an element of the set A. If both a and b are elements of the set A, we may write $a, b \in A$.

If P is the set of all positive integers, $a \in P$ means merely that a is a positive integer. Certaintly, then, it is true that $1 \in P$, $2 \in P$, and so on. If B is the set of all triangles in a given plane, $a \in B$ means that a is one of the triangles in this plane. If C is the set of all books in the Library of Congress, then $a \in C$ means that a is one of these books. We shall presently give other examples of sets.

If $a, b \in A$ and we write $a = b$, *it is always to be understood that these are identical elements of* A. Otherwise expressed, a and b are merely different symbols designating the same element of A. If $a, b \in A$ and it is not true that $a = b$, we may indicate this fact by writing $a \neq b$ and may say that a and b are *distinct* elements of A.

If A and B are sets with the property that every element of A is also an element of B, we call A a *subset* of B and write $A \subseteq B$ (read, "A is contained in B"). An alternate way of expressing the fact that $A \subseteq B$ is to write $B \supseteq A$ (read, "B contains A"). If it is not true that $A \subseteq B$, we may indicate this fact by writing $A \nsubseteq B$. Perhaps we should point out that for every set A it is true that $A \subseteq A$ and hence, according to our definition, one of the subsets of A is A itself. If $A \subseteq B$ and also $B \subseteq A$, then A and B have exactly the same elements and we say that these sets are *equal*, and indicate this by writing $A = B$. If it is not true that $A = B$, we may write $A \neq B$. If $A \subseteq B$ and $A \neq B$, then we say that A is a *proper subset* of B and indicate this fact by the notation $A \subset B$ (read, "A is properly contained in B"). Clearly, $A \subset B$ means that every element of A is an element of B and, moreover, B contains at least one element which is not an element of A.

Sometimes, as has been the case so far, we may specify a set by stating in words just what its elements are. Another way of specifying a set is to exhibit its elements, usually enclosed between braces. Thus, $\{x\}$ indicates the set which consists of the single element x, $\{x, y\}$ the set consisting of the two elements x and y, and so on. We may write $A = \{1, 2, 3, 4\}$ to mean that A is the set whose elements are the positive integers 1, 2, 3, and 4. If P is the set of all positive integers, by writing

$$K = \{a \mid a \in P, a \text{ divisible by } 2\}$$

we shall mean that K consists of all elements a having the properties indicated after the vertical bar, that is, a is a positive integer and is divisible by 2. Hence, K is just the set of all *even* positive integers. We may also write

$$K = \{2, 4, 6, 8, \cdots\},$$

the dots indicating that all even positive integers are included in this set. As another example, if

$$D = \{a \mid a \in P, a < 6\},$$

then it is clear that $D = \{1, 2, 3, 4, 5\}$.

Whenever we specify a set by exhibiting its elements, it is to be understood that the indicated elements are distinct. Thus, for example, if we write $B = \{x, y, z\}$, we mean to imply that $x \neq y$, $x \neq z$, and $y \neq z$.

For many purposes, it is convenient to allow for the possibility that a set may have no elements. This fictitious set with no elements we shall call the *empty set*. According to the definition of subset given above, the empty set is a subset of every set. Moreover, it is a proper subset of every set except the empty set itself. The empty set is often designated by $\varnothing$, and thus we have $\varnothing \subseteq A$ for every set A.

If A and B are sets, the elements that are in both A and B form a set called the *intersection* of A and B, denoted by $A \cap B$. Of course, if A and B have no elements in common, $A \cap B = \varnothing$.

If A and B are sets, the set consisting of those elements which are elements either of A or of B (or of both) is a set called the *union* of A and B, denoted by $A \cup B$.

As examples of the concepts of intersection and union, let $A = \{1, 2, 3\}$, $B = \{2, 4, 5\}$, and $C = \{1, 3, 6\}$. Then we have $A \cap B = \{2\}$, $A \cap C = \{1, 3\}$, $B \cap C = \varnothing$, $A \cup B = \{1, 2, 3, 4, 5\}$, $A \cup C = \{1, 2, 3, 6\}$, and $B \cup C = \{1, 2, 3, 4, 5, 6\}$.

Although we have defined the intersection and the union of only *two* sets, it is easy to extend these definitions to any number of sets, as follows. The *intersection* of any number of given sets is the set consisting of those elements which are in all the given sets and the *union* is the set consisting of those elements which are in at least one of the given sets.

If A, B, and C are sets, each of the following is an almost immediate consequence of the various definitions which we have made:

$A \cap B \subseteq A$ and $A \cap B \subseteq B$.
$A \subseteq A \cup B$ and $B \subseteq A \cup B$.
$A \cap B = A$ if and only if $A \subseteq B$.
$A \cup B = A$ if and only if $B \subseteq A$.
If $B \subseteq C$, then $A \cup B \subseteq A \cup C$ and $A \cap B \subseteq A \cap C$.

In two of these statements, we have used the expression "if and only if." Thus, in accordance with the explanation given in the preceding chapter, we have to establish two different implications. For example, to show that "$A \cap B = A$ if and only if $A \subseteq B$," we need to verify the truth of both of the following implications:

$$\text{If } A \cap B = A, \quad \text{then} \quad A \subseteq B\,.$$
$$\text{If } A \subseteq B, \quad \text{then} \quad A \cap B = A\,.$$

Naturally, we could just as well use the contrapositive form of either or both of these implications. For example, the second one is equivalent to the implication

$$\text{If } A \cap B \neq A, \quad \text{then} \quad A \nsubseteq B\,.$$

We leave to the reader the simple verification of all the implications stated above.

In working with sets, so-called Venn diagrams are sometimes used to give a purely symbolic, but convenient, geometric indication of the relationships involved. Suppose, for the moment, that all sets being considered are subsets of some fixed set U. In Figures 1 and 2,

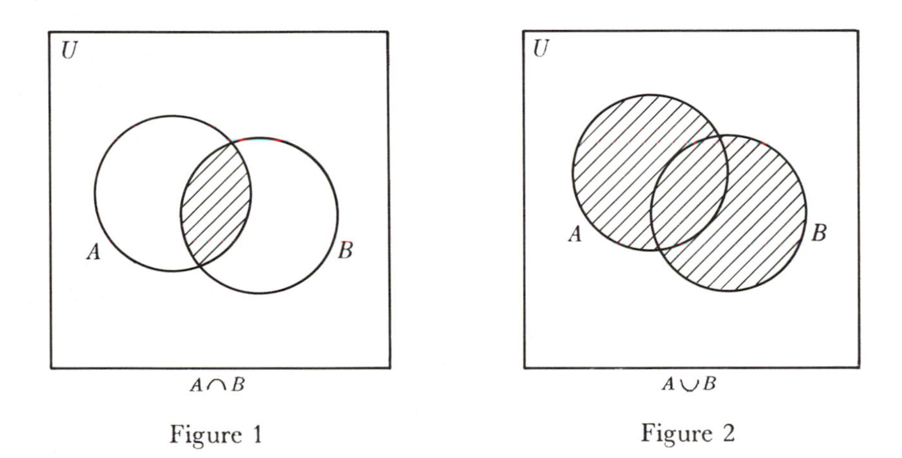

Figure 1

Figure 2

the points within the square represent elements of U. If A and B are subsets of U, then the elements of A and B may be represented by the points within indicated circles (or any other closed regions). The intersection and the union of the sets A and B are then represented in an obvious way by the shaded regions in Figures 1 and 2, respectively.

Of course, the use of a Venn diagram is not meant to imply anything about the nature of the sets being considered, whether or not indicated intersections are nonempty, and so on. Moreover, such a diagram cannot in itself constitute a proof of any fact, but it may be quite helpful in suggesting a proof.

Let us make the following remarks by way of emphasis. A problem of frequent occurrence is that of proving the equality of two sets. Suppose that C and D are given sets and it is required to prove that $C = D$. By definition of equality of sets, we need to show that $C \subseteq D$ and $D \subseteq C$. Sometimes one or both of these conditions follow easily from given facts. If not, the standard procedure is to start with an arbitrary element of C and show that it is an element of D, and then do the same thing with C and D interchanged. When we write "let $x \in C$" or "if $x \in C$," we mean that x is to represent a completely arbitrary element of the set C. Hence, to show that $C \subseteq D$, we only need to show that "if $x \in C$, then $x \in D$." Of course, any other symbol could be used in place of x. Let us now give an example by way of illustration.

Example: If A, B, and C are sets, prove that

$$A \cup (B \cap C) = (A \cup B) \cap (A \cup C).$$

Solution: First, let us take advantage of the opportunity to give another illustration of a Venn diagram. If we think of the meaning of $A \cup (B \cap C)$ as consisting of all elements of A together with all elements that are in both B and C, we see that the set $A \cup (B \cap C)$ may be represented by the shaded portion of the Venn diagram in Figure 3.

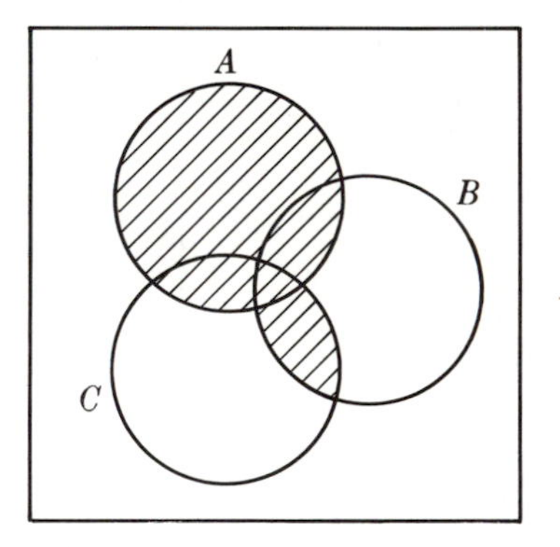

Figure 3

We leave it to the reader to verify that this same shaded region also represents the set $(A \cup B) \cap (A \cup C)$.

We now proceed to give a formal proof of the required formula. Clearly, $B \cap C \subseteq B$, so $A \cup (B \cap C) \subseteq A \cup B$. Similarly, $B \cap C \subseteq C$, and hence $A \cup (B \cap C) \subseteq A \cup C$. It follows that

$$A \cup (B \cap C) \subseteq (A \cup B) \cap (A \cup C),$$

and we have obtained inclusion one way. To obtain inclusion the other way, let $x \in (A \cup B) \cap (A \cup C)$ and let us show that $x \in A \cup (B \cap C)$. Now $x \in A \cup B$ and also $x \in A \cup C$. If $x \in A$, then surely $x \in A \cup (B \cap C)$. If $x \notin A$, then $x \in B$ and $x \in C$, so that $x \in B \cap C$, and again we have that $x \in A \cup (B \cap C)$. This shows that $(A \cup B) \cap (A \cup C) \subseteq A \cup (B \cap C)$, and the proof is therefore complete.

If A and B are subsets of some set U, it is at times convenient to have a notation for the set of elements of A that are *not* elements of B. This set is sometimes denoted by $A - B$, but we shall use the notation $A \setminus B$ in order that there can be no confusion with other uses of the "minus" symbol. For example, if $A = \{1, 2, 3\}$ and $B = \{2, 4\}$, then $A \setminus B = \{1, 3\}$.

Another important concept is illustrated by the familiar idea of coordinates of a point in a plane. A point is determined by an ordered pair (x, y) of real numbers. The word *ordered* is meant to imply that the order of writing the two numbers x and y is important; that is, that (x, y) is to be considered as a different pair than (y, x) unless, of course, x and y happen to be equal real numbers. If $\mathbf{R}$ denotes the set of all real numbers, the set of all ordered pairs of elements of $\mathbf{R}$ is frequently called the *Cartesian product* of $\mathbf{R}$ by $\mathbf{R}$ and designated by $\mathbf{R} \times \mathbf{R}$. More generally, if A and B are any sets, the set of all ordered pairs (a, b), where $a \in A$ and $b \in B$, is the *Cartesian product* of A by B, designated by $A \times B$. It may happen, of course, that A and B are identical sets, as in the illustration given above. It is obvious how to define the Cartesian product of more than two sets. Thus, for example, the set $A \times B \times C$ is the set of all ordered triples (a, b, c), where $a \in A$, $b \in B$, and $c \in C$.

As another example of a product set, if $A = \{1, 2, 3\}$ and $B = \{u, v\}$, then

$$A \times B = \{(1, u), (1, v), (2, u), (2, v), (3, u), (3, v)\}.$$

The final concept to be introduced in this section is a set whose elements are themselves sets; more specifically, the elements are the subsets of some given set. If A is a given set, the set of all subsets of A is often called the *power set* of A and designated by $\mathscr{P}(A)$. For example, if $A = \{1, 2\}$, then the elements of the power set of A are the subsets of A; that is,

$$\mathscr{P}(A) = \{\varnothing, \{1\}, \{2\}, \{1, 2\}\}.$$

EXERCISES

1. If $A = \{a, b, c\}$, $B = \{c, x, y\}$, and $C = \{x, y\}$, determine each of the following sets: $A \cap B$, $A \cap C$, $A \cup B$, $A \cup C$, $A \times C$, $C \times A$, $A \setminus B$, $\mathscr{P}(A)$, $C \times \mathscr{P}(C)$.
2. Let P be the set of all positive integers, and define subsets of P as follows:

$$\begin{aligned} F &= \{a \mid a \in P, a < 10\}, \\ G &= \{a \mid a \in P, a > 5\}, \\ H &= \{a \mid a \in P, a \text{ divisible by } 3\}. \end{aligned}$$

   Determine each of the following sets: $F \cap G$, $F \cap H$, $G \cap H$, $F \cup G$, $F \cup H$, $G \cup H$.
3. If A, B, and C are sets, draw Venn diagrams to illustrate and then give a formal proof that $A \cap (B \cup C) = (A \cap B) \cup (A \cap C)$.
4. If A, B, and C are subsets of some set U, prove each of the following:

   (i) $$A \setminus (B \cup C) = (A \setminus B) \cap (A \setminus C),$$
   (ii) $$A \setminus (B \cap C) = (A \setminus B) \cup (A \setminus C).$$

5. If k is a positive integer, show that a set with $k + 1$ elements has twice as many subsets as a set with k elements.
6. Use the result of the preceding exercise to give a plausible reason (a formal proof is not required) why it is true that if a set A has n elements, where n is a positive integer, then the set $\mathscr{P}(A)$ has 2^n elements.
7. The coefficient of $x^r y^{n-r}$ $(0 \leq r \leq n)$ in the binomial expansion of $(x + y)^n$ is the number of ways in which r objects can be chosen from a set of n objects. Use this fact, applied to $(1 + 1)^n$, to give an alternate proof of the result stated in the preceding exercise.
8. If the set A has n elements, how many elements are there in the set $\mathscr{P}(\mathscr{P}(A))$?
9. If the set A has n elements, n a positive integer, prove that A has as many subsets with an even number of elements as it has subsets with an odd number of elements.

1.2 MAPPINGS

As a first illustration of the concept to be introduced in this section, let C be the set of all books in the Library of Congress and P the set of all positive integers. Corresponding to each book there is a unique positive integer, namely, the number of pages in the book. That is, to each element of C there corresponds in this way a unique element of P. This is an example of a mapping of the set C into the set P. As another illustration, let N be the set of all names occurring in a given telephone directory, and L the set of the twenty-six letters of the alphabet. We may then associate with each name the first letter of the surname, and this then defines a mapping of N into L. Additional examples will be given after the following definition.

1.1 Definition. A *mapping* of a set A into a set B is a correspondence that associates with each element a of A a unique element b of B. The notation $a \to b$ is sometimes used to indicate that b is the element of B that is associated with the element a of A under a given mapping. We may say that *a maps into b* or that b is the *image* of a under this mapping.

In order to avoid some trivial special cases, whenever we consider a mapping of a set A into a set B we shall always assume that the sets A and B are not empty.

Let us now give an example of a mapping of the set $S = \{1, 2, 3, 4\}$ into the set $T = \{x, y, z\}$. To specify such a mapping, we merely need to select an element of T to be the image of each element of S. Thus

1.2 $$1 \to x, \quad 2 \to y, \quad 3 \to x, \quad 4 \to y$$

defines a mapping of S into T in which x is the image of 1, y the image of 2, and so on. Note that although every element of S is required to have an image in T, it need not be true that every element of T occurs as the image of at least one element of S.

Before proceeding, let us observe that a *function*, as the term is often used, is just a mapping of the set $\mathbf{R}$ of all real numbers (or of some subset of $\mathbf{R}$) into the same set $\mathbf{R}$. For example, the function f defined by $f(x) = x^2 + x + 1$ is the mapping $x \to x^2 + x + 1$ which associates with each real number x the real number $x^2 + x + 1$. In this setting, the mapping is denoted by f and the image of the number x under the mapping f by $f(x)$.

Although we are now concerned with arbitrary sets (not just sets of real numbers), the function notation of the preceding paragraph could be, and frequently is, used for mappings. However, we shall adopt an alternate notation which is also of fairly wide use in algebra and which will have some advantages later on. Mappings will henceforth usually be denoted by Greek letters, such as α, β, γ, $\cdots$. If α is a mapping of A into B, the image of an element a of A will be denoted by $a\alpha$. Note that α is here written on the right and without parentheses, rather than on the left as in the more familiar function notation $\alpha(a)$. For example, let β be the mapping 1.2 of S into T. Then, instead of writing 1.2 we might just as well write

1.3 $$1\beta = x, \quad 2\beta = y, \quad 3\beta = x, \quad 4\beta = y\,.$$

Another mapping γ of S into T is defined by

1.4 $$1\gamma = x, \quad 2\gamma = y, \quad 3\gamma = y, \quad 4\gamma = z\,.$$

We shall presently use these mappings to illustrate certain additional concepts.

It is customary to write $\alpha: A \to B$ to indicate that α is a mapping of the set A into the set B. We may also sometimes write $a \to a\alpha$, $a \in A$, to indicate this mapping, it being understood that for each $a \in A$, $a\alpha$ is a uniquely determined element of B. If we have mappings $\alpha: A \to B$ and $\beta: A \to B$, we naturally consider these mappings to be equal, and write $\alpha = \beta$, if and only if $a\alpha = a\beta$ for every $a \in A$. Thus, for the mappings $\beta: S \to T$ and $\gamma: S \to T$ exhibited above, we have $\beta \neq \gamma$ since, for example, $3\beta = x$ and $3\gamma = y$.

We may point out that associated with the mapping 1.3 in a natural way is a unique subset of the product set $S \times T$, namely, the set $\{(1, x), (2, y), (3, x), (4, y)\}$, in which we use, as the second element of a pair, the image of the first element under the mapping β. More generally, if $\alpha: A \to B$ is a mapping of A into B, we may associate with the mapping α the subset of $A \times B$ whose elements are the pairs $(a, a\alpha)$, $a \in A$. Conversely, suppose that T is a subset of $A \times B$ with the following two properties:

(1) For each element a of A, there exists an element of T of the form (a, b); that is, each element a of A is the first element of some ordered pair in T.
(2) If $(a, b) \in T$ and $(a, c) \in T$, then $b = c$.

These two properties merely assure us that if $a \in A$, there exists exactly *one* element b of B such that $(a, b) \in T$. Since the element b of B is uniquely determined in this way by a, we may define a mapping $\alpha: A \to B$ by defining $a\alpha = b$. For this reason, a mapping of A into B is sometimes *defined* to be a subset of $A \times B$ with the two properties stated above. Although we shall not take this viewpoint, it should be observed that this approach makes it possible to define a mapping without using the undefined word *correspondence* which was used in Definition 1.1.

In studying mappings it is sometimes suggestive to make use of a geometrical diagram. Figure 4 is supposed to suggest that α is a

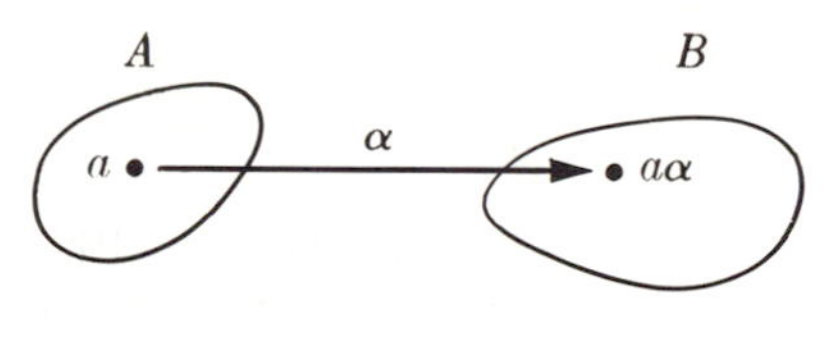

Figure 4

mapping of A into B and that under this mapping each element a of A has image $a\alpha$ in B.

In the particular mapping $\beta: S \to T$ given by 1.3, the element z of T does not occur as the image of any element of S. However, in the mapping $\gamma: S \to T$, defined in 1.4, every element of T is the image of at least one element of S. The language for stating this essential difference between these mappings is given in the following definition.

1.5 Definition. A mapping α of A into B is said to be a mapping of A *onto* B if and only if every element of B is the image of at least one element of A under the mapping α.

Thus, γ is a mapping of S onto T, whereas β is not a mapping of S onto T. It is important to observe that "into" is *not* the opposite of "onto." According to our language, every mapping is a mapping of some set into some set. That is, "onto" is a special case of "into," and if $\alpha: A \to B$ is a mapping of A *onto* B, it is perfectly correct to say that it is a mapping of A into B (although this doesn't give the maximum amount of available information).

If $\alpha: A \to B$ is a mapping of A into B, let us denote by $A\alpha$ the set of all elements of B that occur as images of elements of A under the mapping α, that is,

$$A\alpha = \{a\alpha \mid a \in A\}\,.$$

Thus α is a mapping of A onto B if and only if $A\alpha = B$. In any case, an arbitrary mapping $\alpha: A \to B$ may be considered as defining a mapping of A *onto* the subset $A\alpha$ of B. Thus, associated with each mapping is an onto mapping if we suitably restrict the set in which the images lie.

There is one additional concept which plays an important role in the study of mappings. In the mapping $\beta: S \to T$, defined by 1.3, we see that both 1 and 3 have x as image. Similarly, $\gamma: S \rightarrow T$, defined by 1.4, is such that both 2 and 3 have y as image. Now let $T = \{x, y, z\}$ as above, and let $U = \{r, s, t, u\}$. Then the mapping $\theta: T \rightarrow U$ defined by

1.6 $$x\theta = t, \quad y\theta = r, \quad z\theta = u$$

is such that every element of U which occurs as an image of some element of T is the image of exactly one element of T. This property has a name which we proceed to introduce.

1.7 Definition. A mapping $\alpha: A \rightarrow B$ is said to be a *one-one* mapping of A into B if and only if distinct elements of A have distinct images in B or, equivalently, if $a_1, a_2 \in A$ such that $a_1\alpha = a_2\alpha$, then $a_1 = a_2$.

The mapping $\theta: T \rightarrow U$ defined by 1.6 is an example of a one-one mapping. Note, however, that it is not an onto mapping. Hence a one-one mapping may or may not be an onto mapping. Clearly, also, an onto mapping need not be a one-one mapping.*

We now give some additional examples to illustrate these concepts.

Example 1: Let C be a nonempty subset of the set D. The mapping $\phi: C \rightarrow D$ defined by $c\phi = c$ for each $c \in C$ is a one-one mapping of C into D. It is an onto mapping if and only if $C = D$.

Example 2: Let $\alpha: A \times B \rightarrow A$ be defined by $(a, b)\alpha = a$ for each $(a, b) \in A \times B$. This is certainly an onto mapping. However, if $b_1, b_2 \in B$ with $b_1 \neq b_2$, then $(a, b_1)\alpha = (a, b_2)\alpha$ with $(a, b_1) \neq (a, b_2)$, so the mapping is not a one-one mapping. It will be a one-one mapping if and only if B has exactly one element. The mapping α of this example is called the *projection* of $A \times B$ onto A. Similarly, one can define the projection of $A \times B$ onto B.

Example 3: Let $\mathbf{Z}$ be the set of all integers and $\alpha: \mathbf{Z} \rightarrow \mathbf{Z}$ be defined by $i\alpha = 2i + 1$, $i \in \mathbf{Z}$. In contrast to most of our previous examples, this is an example of a mapping of the set $\mathbf{Z}$ into the same set $\mathbf{Z}$. To determine whether α is an onto mapping, let j be an arbitrary element of $\mathbf{Z}$ and let us find whether j is the image of some element. That is, we need

* Some other terms will be found in the literature as follows. An onto mapping is also called a *surjection;* a one-one mapping may be called an *injection.* A mapping which is both one-one and onto is also called a *bijection.*

to determine whether there exists an integer i such that $2i + 1 = j$. Clearly, there will be no such integer i if j is even since $2i + 1$ is odd for every integer i. Thus, α is not an onto mapping. Is it a one-one mapping? To answer this question, suppose that $i_1, i_2 \in \mathbf{Z}$ such that $i_1\alpha = i_2\alpha$, that is, such that $2i_1 + 1 = 2i_2 + 1$. It follows that $i_1 = i_2$ and α is therefore a one-one mapping.

The mapping of a set A into itself in which each element is its own image is often called the *identity mapping* on the set A. If we denote this identity mapping by ϵ_A, we see that the identity mapping ϵ_A on A is defined by $a\epsilon_A = a$ for each $a \in A$. Thus, the mapping ϕ of Example 1 is the identity mapping on C if and only if $C = D$. It is clear that an identity mapping is always one-one and onto.

We conclude our present discussion of mappings by considering one additional concept as follows. Suppose that $\alpha: A \to B$ is a *one-one* mapping of A *onto* B, that is, it has both the "one-one" and "onto" properties. Then each element of B is expressible in the form $a\alpha$ for exactly one element a of A. We can therefore define in a natural way a mapping of B onto A by making a the image of $a\alpha$ for each $a \in A$. This particular mapping of B onto A is often denoted by α^{-1} (since it reverses the effect of α). That is, the mapping $\alpha^{-1}: B \to A$ is defined by $(a\alpha)\alpha^{-1} = a$, $a \in A$. Clearly, α^{-1} is a one-one mapping of B onto A. The simple relationship between the mappings α and α^{-1} may possibly be suggested by the diagram of Figure 5. The reader should carefully

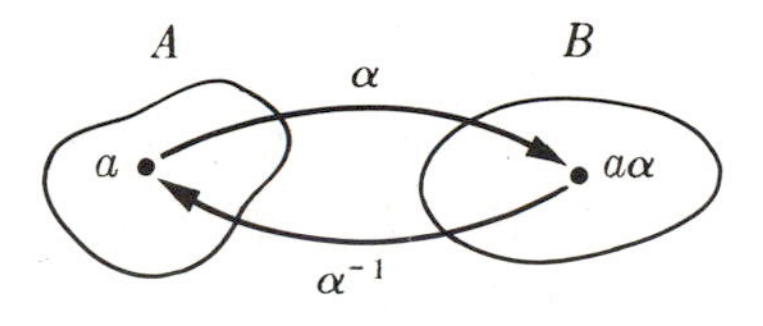

Figure 5

observe that the mapping α^{-1}, as here defined, exists if and only if α is a one-one mapping of A onto B.

Clearly, a one-one mapping of A onto B may be thought of as a pairing of the elements of A and the elements of B. In view of the mutual relationship between A and B, a one-one mapping of A onto B (or of B onto A) is sometimes called a *one-to-one correspondence between A and B*.

It will probably seem reasonable, and it is indeed true although we shall not discuss this fact here, that if set C has n elements for some positive integer n, then there will exist a one-one mapping of C onto D if and only if D also has n elements.

EXERCISES

1. Let $\mathbf{Z}$ be the set of all integers, and $i \in \mathbf{Z}$. Determine in each case whether the indicated mapping α of $\mathbf{Z}$ into $\mathbf{Z}$ is an onto mapping and whether it is a one-one mapping.

   (a) $i\alpha = i + 3$, (b) $i\alpha = i^2 + i$,
   (c) $i\alpha = i^3$, (d) $i\alpha = 2i - 1$,
   (e) $i\alpha = -i + 5$, (f) $i\alpha = i - 4$.

2. Let $\mathbf{R}$ be the set of all real numbers, and $x \in \mathbf{R}$. Determine in each case whether the indicated mapping α of $\mathbf{R}$ into $\mathbf{R}$ is an onto mapping and whether it is a one-one mapping.

   (a) $x\alpha = 2x + 1$, (b) $x\alpha = 1 - x$,
   (c) $x\alpha = x^2$, (d) $x\alpha = x^3$,
   (e) $x\alpha = x^2 + x$, (f) $x\alpha = 4x$,
   (g) $x\alpha = \begin{cases} x \text{ if } x \text{ is rational}, \\ 2x \text{ if } x \text{ is irrational}. \end{cases}$

3. Let P be the set of all positive integers, and $n \in P$. Determine in each case whether the indicated mapping of P into P is an onto mapping and whether it is a one-one mapping.

   (a) $n\alpha = 2n$, (b) $n\alpha = n + 1$,
   (c) $n\alpha = n^2$, (d) $1\alpha = 1$, $n\alpha = n - 1$ for $n > 1$.

4. Give an example of a mapping of the set P of all positive integers into the set P such that every element of P is the image of exactly two elements.

5. If P is as in the preceding exercise, give several examples of mappings $\alpha: P \to P$, other than the identity mapping on P, such that α is a one-one and onto mapping.

6. If $\mathbf{R}$ is the set of all real numbers, use the fact that every cubic equation with real coefficients has a real root to show that the mapping α of $\mathbf{R}$ into $\mathbf{R}$ defined by $x\alpha = x^3 - x$, $x \in \mathbf{R}$, is a mapping of $\mathbf{R}$ onto $\mathbf{R}$. Is it a one-one mapping?

7. If $\mathbf{R}$ is the set of all real numbers, why doesn't the formula $x\alpha = 1/x$, $x \in \mathbf{R}$, define a mapping of $\mathbf{R}$ into $\mathbf{R}$?

8. If $A = \{1, 2, 3\}$ and $B = \{x, y\}$, verify that there exist eight mappings of A into B and nine mappings of B into A. How many mappings of A onto B are there?

9. Let A be a set with m elements and B a set with n elements (m and n positive integers). Formal proofs are not required, but in each of the following give some indication why you believe your conclusion to be correct.

(a) Determine the number of mappings of A into B.

(b) If $n \geq m$, determine the number of one-one mappings of A into B.

(c) If $m = n$, determine the number of one-one mappings of A onto B.

10. Let $\mathbf{R}$ be the set of all real numbers, and let $x \in \mathbf{R}$. Use any calculus methods which you know to determine whether each of the following mappings α of $\mathbf{R}$ into $\mathbf{R}$ is an onto mapping and whether it is a one-one mapping:

(a) $x\alpha = e^x$, (b) $x\alpha = \sin x$,

(c) $x\alpha = x + \sin x$, (d) $x\alpha = \frac{1}{2}x + \sin x$.

1.3 EQUIVALENCE RELATIONS

As a simple illustration of the next concept to be introduced, let the "less than" symbol "$<$" have the usual meaning as applied to integers. Since for every ordered pair (i, j) of integers, $i < j$ is either true or false, we say that "$<$" is a relation *defined* on the set $\mathbf{Z}$ of integers.

In general, let A be a given set. We then say that a *relation* R is defined on A if for each ordered pair (a, b) of elements of A it is true or false that a is in the relation R to b. It is to be understood that if a and b are given, enough information is available to determine whether or not a is in the relation R to b. It is customary to write aRb to indicate that a is in the relation R to b. If, for example, we let R be the relation $<$ on $\mathbf{Z}$, then iRj merely means that $i < j$.

If R is a given relation on a set A, associated with this relation in a uniquely determined manner is a subset of the product set $A \times A$, namely, the subset consisting of all ordered pairs (a, b) such that aRb. Conversely, given a subset R of $A \times A$, we can use it to determine a relation R on A by simply defining aRb to mean that the ordered pair (a, b) is an element of R. For this reason, a relation on a set A is sometimes *defined* to be a subset of $A \times A$.

We shall usually not be concerned with relations in general, but primarily with those relations having the particular properties stated in the following definition.

1.8 Definition. A relation R defined on a set A is called an *equivalence relation* if it has the following three properties, where a, b, and c are arbitrary elements of A:

(1) aRa (*reflexive property*).
(2) If aRb, then bRa (*symmetric property*).
(3) If aRb and bRc, then aRc (*transitive property*).

In the future, we shall usually use "$\sim$" to denote an equivalence relation. It may then be convenient to read $a \sim b$ as "a is equivalent to b." We shall sometimes write $a \nsim b$ to indicate that a is not equivalent to b.

Let us emphasize how property (1) above differs in an essential way from the other two properties. If we have an equivalence relation "$\sim$" defined on a set A, then property (1) asserts that $a \sim a$ for *every* element a of A. On the other hand, if a and b are given, property (2) says nothing about whether it is true that $a \sim b$, only that *if* it is true, then we must also have $b \sim a$. A similar remark holds for the transitive property. In other words, both the symmetric property and the transitive property assert the truth of an implication.

The relation "$<$" on the set **Z** of all integers is not an equivalence relation since it has neither the reflexive property nor the symmetric property. The relation "$\leq$" has the reflexive property and the transitive property, but not the symmetric property. Of course, "$=$" is an equivalence relation on **Z** (or on any other set), as is perhaps suggested by the word "equivalence."

We shall now give a few examples of equivalence relations, but many more will occur in later chapters of this book. Let T be the set of all triangles in a fixed plane, and let a and b be arbitrary elements of T. Then "$\sim$" is an equivalence relation on T if we agree to define "$\sim$" in any one of the following ways:

(i) $a \sim b$ to mean "a is congruent to b,"
(ii) $a \sim b$ to mean "a is similar to b,"
(iii) $a \sim b$ to mean "a has the same area as b,"
(iv) $a \sim b$ to mean "a has the same perimeter as b."

As another example of an equivalence relation, let **Z** be the set of all integers, and let us define $a \equiv b$ to mean that $a - b$ has 3 as a factor; that is, that there exists an integer n such that $a - b = 3n$. It is then readily verified that "$\equiv$" has the three defining properties of an equivalence relation. Furthermore, every integer is equivalent to one of the three integers 0, 1, 2. In this connection, consider the following three subsets of **Z**:

$$\begin{aligned} J &= \{\cdots, -9, -6, -3, 0, 3, 6, 9, \cdots\}, \\ K &= \{\cdots, -8, -5, -2, 1, 4, 7, 10, \cdots\}, \\ L &= \{\cdots, -7, -4, -1, 2, 5, 8, 11, \cdots\}, \end{aligned}$$

It will be observed that every integer is in exactly one of these subsets. In other words, the union of these three subsets is **Z** and the intersection of any two of them is the empty set. Moreover, J can be characterized as the set of all elements of **Z** that are equivalent to 0 (or to any other element of J), and similar characterizations can be given for K and L. The sets J, K, and L are examples of a concept which we proceed to define.

1.9 Definition. Let A be a set and "$\sim$" an equivalence relation defined on A. If $a \in A$, the subset of A which consists of all elements x of A such that $x \sim a$ is called an *equivalence set*. This equivalence set will frequently be denoted by $[a]$.

This definition of the equivalence set $[a]$ can be written formally as follows:

1.10
$$[a] = \{x \mid x \in A, x \sim a\}.$$

In the above example, note that $J = [0]$, $K = [1]$, and $L = [2]$; also that $[0] = [3] = [6]$, and so on. Hence there are just the three different equivalence sets.

To return to the general definition, let us consider a few properties of equivalence sets. First, since $a \sim a$ by the reflexive property of an equivalence relation, we always have $a \in [a]$; that is, $[a]$ is the equivalence set which contains a. This shows that every element of A is in at least one equivalence set. Other important properties of equivalence sets are the following, where a and b are elements of the set A:

1.11
(i) $[a] = [b]$ if and only if $a \sim b$;
(ii) if $[a] \cap [b] \neq \varnothing$, then $[a] = [b]$.

As a first step in proving 1.11(i), let us assume that $[a] = [b]$ and show that $a \sim b$. It has been pointed out that $a \in [a]$, and hence we have $a \in [b]$. By definition of the equivalence set $[b]$, it follows that $a \sim b$, as we wished to show. Conversely, let us now assume that $a \sim b$. If $x \in [a]$, then $x \sim a$ by definition of $[a]$. Now we have $x \sim a$

and $a \sim b$, so the transitive property of an equivalence relation assures us that $x \sim b$. This then implies that $x \in [b]$, and we have therefore proved that $[a] \subseteq [b]$. We leave as an exercise the similar proof that $[b] \subseteq [a]$, from which we conclude that $[a] = [b]$, as desired.

We now prove 1.11(ii). Since $[a] \cap [b] \neq \varnothing$, there exists at least one element s of A such that $s \in [a]$ and also $s \in [b]$. It follows that $s \sim a$ and $s \sim b$. (Why?) By the symmetric property of an equivalence relation, we have $a \sim s$. Since $a \sim s$ and $s \sim b$, the transitive property assures us that $a \sim b$. The fact that $[a] = [b]$ then follows at once from 1.11(i).

A collection of nonempty subsets of a set A is often called a *partition* of A if A is the union of these subsets and any two of the subsets have empty intersection. In view of 1.11(ii), together with the fact that every element of A is in some equivalence set, we see that the different equivalence sets relative to an equivalence relation defined on A form a partition of A. Conversely, suppose that we are given a partition of the set A. If $a, b \in A$, and we define "$\sim$" by asserting that $a \sim b$ if and only if a and b are elements of the same subset of A in the given partition, then it is easy to verify that "$\sim$" is an equivalence relation; clearly the equivalence sets are the different subsets of A in the given partition. It follows that there is a one-to-one correspondence between equivalence relations on A and partitions of A.

1.4 OPERATIONS

There is just one other term that we wish to introduce in this preliminary chapter. First, we consider a familiar concept as follows. Let **Z** be the set of all integers. Associated with each ordered pair (i, j) of elements of **Z** there is a uniquely determined element $i + j$ of **Z**. Accordingly, we say that addition, denoted by "$+$," is an operation on **Z**. More precisely, we may call it a *binary* operation to emphasize that it is defined for each ordered *pair* of elements of **Z**. The general definition is as follows.

1.12 Definition.* Let A be a given set. *A binary operation* "$\circ$" on A is a correspondence that associates with each ordered pair (a, b) of elements of A a uniquely determined element $a \circ b$ of A.

* Expressed more formally, this definition merely asserts that a binary operation "$\circ$" on A is a mapping of $A \times A$ into A. The image of an element (a, b) of $A \times A$ under this mapping is then denoted by $a \circ b$.

It may be emphasized that if a binary operation "$\circ$" is defined on the set A, this is supposed to imply that $a \circ b$ must always be an element *of the set* A, for $a, b \in A$. In this connection, suppose that B is a subset of A and that we have an operation "$\circ$" defined on A. If it happens that for $x, y \in B$ it is always true that $x \circ y$ (which we know is an element of A) is actually an element of the subset B, then it is customary to say that B is *closed* under (or with respect to) the operation "$\circ$." Of course, in this case, we may consider "$\circ$" to be an operation on the set B. As examples, consider ordinary addition and multiplication defined on the set $\mathbf{Z}$ of all integers, and let S be the set of all *odd* integers. Then we see that S is closed under multiplication, but is not closed under addition. The set of all *even* integers is closed under both addition and multiplication.

Later on we shall seldom have occasion to use any unfamiliar symbol to denote an operation. For the most part, we shall find it convenient to call an operation "addition" or "multiplication," and to use the familiar notations $a + b$ and $a \cdot b$ (or simply ab). However, for the moment we continue to use the symbol "$\circ$" for a binary operation on a set A. We may emphasize that saying that "$\circ$" is a binary operation on A asserts that $a \circ b$ is a uniquely determined element of A for *every* $a \in A$ and *every* $b \in A$. Some important concepts are introduced in the following definition.

1.13 Definition. Let "$\circ$" be a binary operation defined on the set A. Then

(i) The operation "$\circ$" is said to be a *commutative* operation if and only if $a \circ b = b \circ a$ for all $a, b \in A$.
(ii) The operation "$\circ$" is said to be an *associative* operation if and only if $(a \circ b) \circ c = a \circ (b \circ c)$ for all $a, b, c \in A$.
(iii) An element e of A is said to be an *identity* for the operation "$\circ$" if and only if $a \circ e = e \circ a = a$ for every $a \in A$.

As examples of these concepts, let us again consider the set $\mathbf{Z}$ of all integers. For the present, we assume as known the familiar properties of addition and multiplication on $\mathbf{Z}$, in particular, that they are both commutative and associative. Moreover, since $a + 0 = 0 + a = a$ for every $a \in \mathbf{Z}$, we see that 0 is the identity for addition; and clearly 1 is the identity for multiplication.

On the same set $\mathbf{Z}$, let us define $a \circ b = a - b$. Since $3 \circ 2 = 1$ and $2 \circ 3 = -1$, we see that this operation is not commutative. Note that just *one* instance in which $a \circ b \neq b \circ a$ implies that the operation is not commutative. The reader may verify that neither is this operation associative. Does there exist an identity for this operation? Since $a \circ 0 = a - 0 = a$ for every integer a, it might appear at first glance that 0 is an identity. However, $0 \circ a = -a$ and the definition of an identity is not met.

EXERCISES

1. On the set of all nonempty subsets of a nonempty set A, consider the relation R defined by aRb if and only if $a \cap b \neq \varnothing$, Which of the three defining properties 1.8 of an equivalence relation hold for this relation?

2. Let R be the relation on the set $\{1, 2\}$ defined by $1R1$ and the relation R holds for no other ordered pair except the pair $(1, 1)$. Show that the relation R has exactly two of the defining properties of an equivalence relation.

3. Give an example of a relation R on some set such that R has the symmetric property but does not have the reflexive property or the transitive property.

4. If "$\sim$" is an equivalence relation on a set A, carefully prove each of the following:

(i) If $a, b \in A$ such that $a \nsim b$, then $[a] \cap [b] = \varnothing$.

(ii) If $a, b, c, d \in A$ such that $c \in [a]$, $d \in [b]$, and $[a] \neq [b]$, then $c \nsim d$.

5. If a and b are integers, let us define $a \equiv b$ to mean that $a - b$ has 5 as a factor. Verify that "$\equiv$" is an equivalence relation on the set $\mathbf{Z}$ of all integers, and exhibit all the different equivalence sets.

6. In each of the following, "$\circ$" is the specified binary operation on the set $\mathbf{Z}$ of integers. Determine in each case whether the operation is commutative, whether it is associative, and whether there is an identity for the operation.

(i) $a \circ b = b$, (ii) $a \circ b = a + b + ab$,

(iii) $a \circ b$ is the larger of a and b, (iv) $a \circ b = 2a + 2b$,

(v) $a \circ b = a + b - 1$, (vi) $a \circ b = a + ab$.

7. Let $\mathscr{P}(A)$ be the power set of a set A.

(i) Is the binary operation "$\cap$" on $\mathscr{P}(A)$ commutative? Is it associative? Does it have an identity?

(ii) Answer the same questions for the binary operation "$\cup$" on $\mathscr{P}(A)$.

Chapter 2

Rings

In this chapter we shall introduce the important class of algebraic systems that are called *rings*, give a large number of examples, and then establish some fundamental properties of any ring. All the properties that are used to define a ring are suggested by simple properties of the integers, and we begin by pointing out some of these properties. The following section is therefore of a preliminary nature and is merely intended to furnish a partial motivation of the material to follow.

2.1 FORMAL PROPERTIES OF THE INTEGERS

The simplest numbers are the numbers 1, 2, 3, $\cdots$, used in counting. These are called the "natural numbers" or the "positive integers." Addition and multiplication of natural numbers have simple interpretations if we consider a natural number as indicating the number of elements in a set. For example, suppose that we have two piles of stones, the first one containing m stones and the second one n stones. If the stones of the first pile are placed on the second pile, there results a pile of $n + m$ stones. If, instead, the stones of the second pile are placed on the first pile, we get a pile of $m + n$ stones. It thus seems quite obvious that

$$m + n = n + m$$

for every choice of m and n as natural numbers, that is, that addition of natural numbers is commutative. This property of the natural numbers is an example of what is sometimes called a *law* or a *formal property*. Another example is the associative law of addition.

Multiplication of natural numbers may be introduced as follows. If one has m piles, each of which contains n stones, and all the stones are placed in one pile, the resulting pile will contain mn stones. It is also a familiar fact that multiplication of natural numbers is both commutative and associative. Moreover, addition and multiplication are such that the so-called *distributive law* holds:

$$m(n + k) = mn + mk,$$

where m, n, and k are arbitrary natural numbers.

Historically, the natural numbers were no doubt used for centuries before there was any consideration of their formal properties. However, in modern algebra it is precisely such formal properties that are of central interest. Some of the reasons for this changed viewpoint will become evident later on in this chapter as well as in succeeding chapters.

Of course, if m and n are natural numbers, there need not be a natural number x such that $m + x = n$. In order to be able to solve all equations of this kind, we need to have available the negative integers and zero along with the positive integers. The properties with which we shall be concerned in the next section are suggested by well-known properties of the system of *all* the integers (positive, negative, and zero). Near the end of the next chapter we shall be ready to give what may be called a characterization of the system of all integers although, for the most part, we shall merely assume a familiarity with the simpler properties of this system. In later chapters, the other number systems of elementary algebra will be discussed in some detail. However, even before they are presented in a logical way we shall not hesitate to illustrate parts of our general theory by examples from these familiar number systems.

2.2 DEFINITION OF A RING

The concepts to be presented in this section are of fundamental importance, although a full realization of their generality will probably not become apparent until the examples of the following section are carefully studied.

We begin with a nonempty set R on which there are defined two binary operations, which we shall call "addition" and "multiplication," and for which we shall use the familiar notation. Accordingly, if a, $b \in R$, then $a + b$ and ab (or $a \cdot b$) are uniquely determined elements of the set R. By way of emphasis, we may state this fact in another way by saying that R is to be *closed* under the binary operations which we are calling addition and multiplication. We now assume the following properties or laws, in which a, b, and c are arbitrary elements, distinct or identical, of R.

P_1: $a + b = b + a$ (*commutative law of addition*).

P_2: $(a + b) + c = a + (b + c)$ (*associative law of addition*).

P_3: There exists an element 0 of R such that $a + 0 = a$ for every element a of R (*existence of a zero*).

P_4: If $a \in R$, there exists $x \in R$ such that $a + x = 0$ (*existence of additive inverses*).

P_5: $(ab)c = a(bc)$ (*associative law of multiplication*).

P_6: $a(b + c) = ab + ac$, $(b + c)a = ba + ca$ (*distributive laws*).

Under all these conditions R is said to be a *ring*. Let us repeat this definition in the following formal way.

> **2.1 Definition.** If R is a nonempty set on which there are defined binary operations of addition and multiplication such that Properties P_1–P_6 hold, we say that R is a *ring* (with respect to these definitions of addition and multiplication).

Let us make a few remarks about the defining properties of a ring. First, we may emphasize that we should not think of the elements of a ring as necessarily being *numbers*. Moreover, addition and multiplication are not assumed to have any properties other than those specified. The element "0," whose existence is asserted in P_3, and which we call a *zero*, is actually an identity for the operation of addition since, by P_1, $0 + a = a + 0$ and therefore we also have $0 + a = a$. We do not assume that there is only *one* identity for addition, but later on we shall prove this to be true. Again, we should not think of 0 as being the familiar *number* zero; it is merely an identity for the operation of addition. Finally, we point out that P_4 does not assert that there is only *one* $x \in R$ such that $a + x = 0$, but this fact will also be proved eventually.

All the properties used to define a ring are certainly familiar properties of the integers. Hence, with the usual definitions of addition and multiplication, the set of all integers is a ring. Henceforth, this ring will be denoted by **Z**. For this ring, the zero whose existence is asserted in P_3 is the familiar number zero.

Now let E be the set of all *even* integers (positive, negative, and zero). Using, of course, addition and multiplication as already defined in **Z**, we see that the sum of two elements of E is an element of E, and similarly for the product of two elements. That is, E is closed under the operations of addition and multiplication. Properties P_1, P_2, P_5, and P_6 hold in E since they hold in the larger set **Z**. Moreover, it is clear that P_3 and P_4 also hold in E, and therefore E is itself a ring.

If all elements of a ring S are contained in a ring R, it is natural to call S a *subring* of R. It is understood that addition and multiplication of elements of S are to coincide with addition and multiplication of these elements considered as elements of the larger ring R. Naturally, a set S of elements of R cannot possibly be a subring of R unless S is closed under the operations of addition and multiplication on R since, otherwise, we would not have operations *on the set* S. We see that E, as defined above, is a subring of the ring **Z**. However, the set of all odd integers cannot be a subring of **Z** since this set is not closed under addition; that is, the sum of two odd integers is not always (in fact, is never) an odd integer.

It is important to observe that the definition of a ring does not require that the operation of multiplication be commutative. However, we shall frequently want to consider this property, so let us give it a number as follows:

P_7: If $a, b \in R$, then $ab = ba$ (*commutative law of multiplication*).

A ring which has property P_7 is called a *commutative* ring. If P_7 does not hold, that is, if there exist at least two elements c and d of R such that $cd \neq dc$, then R is said to be a *noncommutative* ring.

We may also point out that in a ring there need not be an identity for the operation of multiplication. If in a ring R there exists an identity for multiplication, we shall usually call it a *unity* of R and say that R is a *ring with unity*. For convenience of reference, let us give this property a number as follows:

P_8: There exists an element e of R such that $ea = ae = a$ for every element a of R (*existence of a unity*).

We may emphasize that a ring need not have either of the properties P_7 or P_8. However, many of the rings that we shall study in detail will have both of these properties. The ring **Z** is an example of a commutative ring with unity, whereas the ring E of all even integers is a commutative ring without a unity. A few cases of noncommutative rings will occur among the examples of the next section. Naturally, they will have to be quite different from the familiar number systems.

2.3 EXAMPLES OF RINGS

In order to give an example of a ring R, it is necessary to specify the elements of R and to define the operations of addition and multiplication on R so that Properties P_1–P_6 hold. The ring **Z** of integers has been mentioned as a well-known example of a ring. Other examples are the ring of all real numbers and the ring of all complex numbers, with the usual definitions of addition and multiplication. It will be recalled that the *rational numbers* are those numbers which can be expressed in the form m/n, where m and n are integers with $n \neq 0$. With respect to the familiar definitions of addition and multiplication of rational numbers, the set of all rational numbers is also a ring. Clearly, the ring **Z** is a subring of the ring of all rational numbers; the ring of all rational numbers is a subring of the ring of all real numbers; and the ring of all real numbers is a subring of the ring of all complex numbers. All these number systems will be considered in detail in later chapters.

We proceed to give some other, less familiar, examples of rings. For the most part, we shall not write out the verifications of the Properties P_1–P_6. Some of these verifications will be required in the next list of exercises. The purpose of these examples is to clarify the concept of a ring and to show that there are rings of many different kinds.

Example 1: Let S be the set of all real numbers of the form $x + y\sqrt{2}$, where $x, y \in \mathbf{Z}$, with addition and multiplication defined in the usual way. It may be verified that S is closed under these operations. Actually, S is a commutative ring with unity. Of course, it is a subring of the ring of all real numbers.

Example 2: Let T be the set of all real numbers of the form $u + v\sqrt[3]{2} + w\sqrt[3]{4}$, where u, v, and w are rational numbers. Using the usual definitions of addition and multiplication, T is a commutative ring with unity.

Example 3: Let $R = \{u, v, w, x\}$; that is, R consists of just these four elements. We define addition and multiplication in R by means of the following tables.

$(+)$	u	v	w	x
u	u	v	w	x
v	v	u	x	w
w	w	x	u	v
x	x	w	v	u

$(\cdot)$	u	v	w	x
u	u	u	u	u
v	u	v	w	x
w	u	w	w	u
x	u	x	u	x

These we read as follows. For example, we find $v + x$ by looking in the addition table at the intersection of the row which contains v as its left-hand element and the column which contains x at the top. Since w appears in this position, we have $v + x = w$. Other examples are: $w + w = u$, $x + w = v$, $vw = w$, $xx = x$. It would take too much calculation to verify the associative laws and the distributive laws, and we shall now merely state that they do hold. From the addition table, it is seen that the zero of the ring R is the element u; and from the multiplication table it follows that v is the unity. The reader may verify that this is a commutative ring. This ring R differs from previous examples in that it has only a finite number (four) of elements.

Example 4: Let C be the set of all functions which are continuous on the closed interval $0 \leq x \leq 1$, with the usual definitions of addition and multiplication of functions. Since a sum or product of two continuous functions is a continuous function, C is closed under these operations. It can be shown that C is a ring. What is the zero of C? Does it have a unity?

Example 5: The set $T = \{0, e\}$ is a ring of two elements if addition and multiplication are defined by the following tables.

$(+)$	0	e
0	0	e
e	e	0

$(\cdot)$	0	e
0	0	0
e	0	e

Clearly, 0 is the zero of this ring and e is the unity. Hence, this ring has *only* a zero and a unity.

Example 6: Let $K = \{a, b, c, d\}$, with addition and multiplication defined by the following tables.

$(+)$	a	b	c	d
a	a	b	c	d
b	b	a	d	c
c	c	d	a	b
d	d	c	b	a

$(\cdot)$	a	b	c	d
a	a	a	a	a
b	a	b	c	d
c	a	a	a	a
d	a	b	c	d

The ring K is our first example of a noncommutative ring. From the multiplication table we see, for example, that $cd = a$, whereas $dc = c$. Does this ring have a unity? What is the zero?

We may emphasize that in this example, as in others in which addition and multiplication of more than two elements are defined by tables, it would be exceedingly tedious to verify the associative and distributive laws. Of course, the tables have not been written down at random but have been obtained by methods not yet available to the student. At present, the associative and distributive laws will have to be taken on faith, but there is no real difficulty in verifying the other defining properties of a ring.

Example 7: For later reference, we give still another example of a ring with four elements a, b, c, and d. In this case, we define addition and multiplication as follows:

$(+)$	a	b	c	d
a	a	b	c	d
b	b	a	d	c
c	c	d	a	b
d	d	c	b	a

$(\cdot)$	a	b	c	d
a	a	a	a	a
b	a	b	c	d
c	a	c	d	b
d	a	d	b	c

It will be observed that the addition table coincides with the addition table of the preceding example. However, the multiplication table is quite different. This ring is another example of a commutative ring.

Example 8: Let L be the set $\mathbf{Z} \times \mathbf{Z} \times \mathbf{Z}$. That is, L is the set of all ordered triples (a, b, c), where $a, b, c \in \mathbf{Z}$. We make the following definitions:

$$(a, b, c) + (d, e, f) = (a + d, b + e, c + f),$$
$$(a, b, c)(d, e, f) = (ad, bd + ce, cf).$$

To avoid any possible confusion, we may again state that we consider two elements of a set to be equal only if they are identical. Hence, if (a, b, c) and (d, e, f) are elements of L, then $(a, b, c) = (d, e, f)$ means that $a = d$, $b = e$, and $c = f$.

It is easy to verify that $(0, 0, 0)$ is the zero of the ring L, and that $(1, 0, 1)$ is a unity. This is another noncommutative ring since, for example,

$$(0, 1, 0)(1, 0, 0) = (0, 1, 0),$$

whereas

$$(1, 0, 0)(0, 1, 0) = (0, 0, 0).$$

Let us verify one of the distributive laws for this ring. If (a, b, c), (d, e, f), and (g, h, i) are elements of L, let us show that

$$(a, b, c)((d, e, f) + (g, h, i)) = (a, b, c)(d, e, f) + (a, b, c)(g, h, i).$$

The equality of these expressions is a consequence of the following simple calculations:

$$\begin{aligned}(a, b, c)((d, e, f) + (g, h, i)) &= (a, b, c)(d + g, e + h, f + i)\\ &= (a(d + g), b(d + g) + c(e + h), c(f + i))\end{aligned}$$

and

$$\begin{aligned}&(a, b, c)(d, e, f) + (a, b, c)(g, h, i)\\ &\qquad = (ad, bd + ce, cf) + (ag, bg + ch, ci)\\ &\qquad = (ad + ag, (bd + ce) + (bg + ch), cf + ci).\end{aligned}$$

The right sides of these equations are equal in view of certain simple properties of the integers. What properties are involved?

Example 9: Let W be the set of all symbols of the form

$$\begin{bmatrix} a & b \\ c & d \end{bmatrix},$$

where a, b, c, and d are arbitrary elements of $\mathbf{Z}$. Our definitions of addition and multiplication are as follows:

$$\begin{bmatrix} a & b \\ c & d \end{bmatrix} + \begin{bmatrix} e & f \\ g & h \end{bmatrix} = \begin{bmatrix} a + e & b + f \\ c + g & d + h \end{bmatrix},$$

$$\begin{bmatrix} a & b \\ c & d \end{bmatrix} \cdot \begin{bmatrix} e & f \\ g & h \end{bmatrix} = \begin{bmatrix} ae + bg & af + bh \\ ce + dg & cf + dh \end{bmatrix}.$$

With respect to these definitions of addition and multiplication, W is a ring. It is called the *ring of all matrices of order two over the integers*. The reader may verify, by examples, that the commutative law of multiplication does not hold and hence that W is a noncommutative ring.

We may point out that the elements of W are quadruples of elements of $\mathbf{Z}$, and could just as well have been written in the form

(a, b, c, d). However, the above notation is more convenient and is the traditional one.

If we modify this example by letting a, b, c, and d be rational (or real, or complex) numbers instead of integers, we obtain *the ring of all matrices of order two over the rational (or real or complex) numbers.*

Example 10: This final example is of a type quite different from any of the previous examples. Let A be a given set, and let R be the power set of A, that is, the set of all subsets of A, including the empty set and the entire set A. We shall now denote elements of R by lower-case letters—even though they are sets of elements of A.

Our definitions of addition and multiplication are as follows. If $a, b \in R$, we define $a + b = (a \cup b) \setminus (a \cap b)$ and $ab = a \cap b$. Note that $a + b$ consists of the elements of A that are in either subset a or in subset b, but not in both. Thus $a + b$ is not, in general, the union of a and b, but it will be this union whenever $a \cap b = \varnothing$. (See Exercise 9

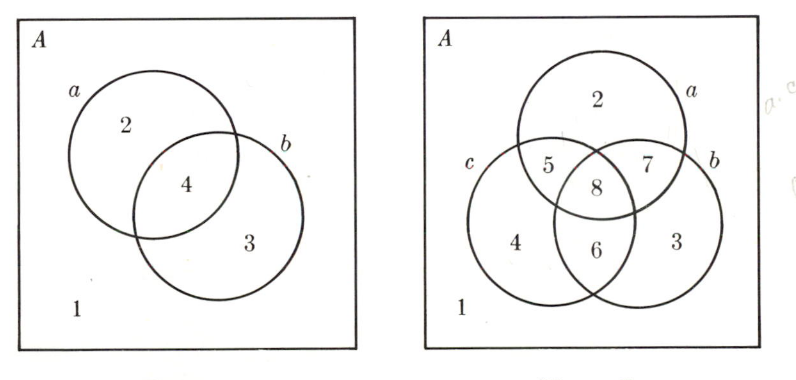

Figure 6 Figure 7

on p. 36.) In the Venn diagram shown in Figure 6, region 1 represents those elements of A which are in neither subset a nor subset b, region 2 represents those elements of A which are in a but not in b, and so on. Hence ab is represented by region 4 and $a + b$ by regions 2 and 3.

We now assert that with the above definitions of addition and multiplication, R is a commutative ring with unity.

The commutative laws of addition and multiplication are obvious, as is also the associative law of multiplication. Let us briefly consider the associative law of addition, and let a, b, and c be arbitrary elements of R. In Figure 7, $a + b$ is represented by regions 2, 3, 5, and 6. Since c is made up of regions 4, 5, 6, and 8, it follows that $(a + b) + c$ is represented by regions 2, 3, 4, and 8. This pictorial representation suggests that $(a + b) + c$ consists of those elements of A which are in exactly one of the subsets a, b, and c together with those which are in all three. To complete the verification of the associative law of addition by means of Venn diagrams, we need to characterize the set $a + (b + c)$. We omit the details, but it is not difficult to verify that we again get the set represented by regions 2, 3, 4, and 8. Hence, $(a + b) + c = a + (b + c)$, as we wished to show. In an exercise below the reader is asked to consider how one could turn this geometrical argument into a formal proof.

If we denote the empty set by "0," it follows that $a + 0 = a$, and the empty set is the zero of the ring R. Moreover, the subset of A consisting of A itself is the unity of the ring. (Why?) If $a \in R$, it is interesting to observe that $a + a = 0$, and thus a is its own additive inverse. Another unusual property of this ring is that $a \cdot a = a$ for every element a of R. We shall refer to this ring as the *ring of all subsets of the set* A. We may emphasize that later on whenever we mention the ring of all subsets of a set *it is always to be understood that addition and multiplication are defined as in this example.*

We conclude this section not by giving still another example of a ring but by presenting a simple, but quite useful, way to construct new rings from given rings. Suppose that R and S are rings, distinct or identical, and let us consider the Cartesian product $R \times S$ whose elements are the ordered pairs (r, s), $r \in R$, $s \in S$. On this set $R \times S$, we define addition and multiplication as follows:

$$(r_1, s_1) + (r_2, s_2) = (r_1 + r_2, s_1 + s_2),$$
$$(r_1, s_1)(r_2, s_2) = (r_1 r_2, s_1 s_2).$$

It is understood, of course, that $r_1, r_2 \in R$ and that $s_1, s_2 \in S$. Moreover, although the same symbol for addition is used in both rings, $r_1 + r_2$ is the sum of r_1 and r_2 in the ring R and $s_1 + s_2$ is the sum of s_1 and s_2 in the ring S (and similarly for products). We leave as an exercise the proof that with respect to the above definitions the set $R \times S$ becomes a ring. It is convenient to have a name for the ring obtained in this way. Accordingly, we make the following definition.

2.2 Definition. If R and S are given rings, the ring whose elements are the elements of the product set $R \times S$, with addition and multiplication as defined above, is called the *direct sum* of the rings R and S, and is usually denoted by $R \oplus S$.

What conditions on R and S will assure us that $R \oplus S$ is commutative? That it has a unity?

That both R + S are commutative

EXERCISES

In these exercises, it is to be assumed that the real numbers (in particular, the rational numbers and the integers) have all the familiar properties which are freely used in elementary algebra.

1. Which of the following are rings with respect to the usual definitions of addition and multiplication? In this exercise, the ring of all even integers is denoted by E.
   (a) The set of all positive integers.
   (b) The set of all integers (positive, negative, and zero) that are divisible by 3.
   (c) The set of all real numbers of the form $x + y\sqrt{2}$, where $x, y \in E$.
   (d) The set of all real numbers of the form $x + y\sqrt[3]{2}$, where $x, y \in \mathbf{Z}$.
   (e) The set of all real numbers of the form $x + y\sqrt[3]{2} + z\sqrt[3]{4}$, where $x, y, z \in \mathbf{Z}$.
   (f) The set of all real numbers of the form $x + y\sqrt{3}$, where $x \in E$ and $y \in \mathbf{Z}$.
   (g) The set of all rational numbers that can be expressed in the form m/n, where $m \in \mathbf{Z}$ and n is a positive odd integer.
2. What is the additive inverse of each element of the ring R of Example 3?
3. Verify that the subset $S = \{u, w\}$ of the ring R of Example 3 is a subring of R. Show that, except for the notation employed, this is the ring of Example 5.
4. For the ring R of Example 3, use the tables to verify each of the following:

$$
\begin{aligned}
(u + v) + w &= u + (v + w), \\
(v + w) + x &= v + (w + x), \\
w(v + x) &= wv + wx, \\
(w + v)x &= wx + vx, \\
(xv)w &= x(vw).
\end{aligned}
$$

5. For the ring L of Example 8, verify the other distributive law and the associative law of multiplication.
6. For the ring W of Example 9, verify the associative law of multiplication and the distributive laws. What is the zero of this ring? Verify that

$$
\begin{bmatrix} 1 & 0 \\ 0 & 1 \end{bmatrix}
$$

is a unity of W. Give examples to show that W is a noncommutative ring.

7. For the ring R of Example 10, consider how a formal proof of the associative law of addition could be given without use of Venn diagrams, and write out at least a part of the proof.

8. For the ring R of Example 10, use Venn diagrams to verify that if $a, b, c \in R$, then $a(b + c) = ab + ac$. How do you know without further calculation that the other distributive law must also hold?

9. In Example 10, why would we not obtain a ring if we defined $ab = a \cap b$ and $a + b = a \cup b$?

10. On the set $S = \mathbf{Z} \times \mathbf{Z}$, let us define addition and multiplication as follows:

$$\begin{aligned}(a, b) + (c, d) &= (a + c, b + d),\\ (a, b)(c, d) &= (ac + 2bd, ad + bc).\end{aligned}$$

Prove that S is a commutative ring with unity.

11. It can be shown that the set $\{a, b, c, d\}$ is a ring if addition and multiplication are defined by the following tables.

$(+)$	a	b	c	d
a	a	b	c	d
b	b	c	d	a
c	c	d	a	b
d	d	a	b	c

$(\cdot)$	a	b	c	d
a	a	a	a	a
b	a	c	a	c
c	a	a	a	a
d	a	c	a	c

Is this a commutative ring? Does it have a unity? What is the zero of this ring? What is the additive inverse of each element of this ring?

12. Show that neither of the following can possibly be the addition table for a ring consisting of the set $\{a, b, c, d\}$ of four elements.

$(+)$	a	b	c	d
a	a	b	c	a
b	b	c	d	a
c	c	d	a	b
d	a	a	b	c

$(+)$	a	b	c	d
a	a	b	c	d
b	b	c	a	d
c	c	a	d	b
d	b	c	d	a

13. Define addition of integers in the usual way, but define the "product" of any two integers to be zero. Is the set of all integers a ring with respect to addition and this new "multiplication"?

14. If a and b are integers, let us define $a \oplus b$ to be ab, and a $\odot$ b to be $a + b$. Is the set of all integers a ring with respect to the operations "$\oplus$" of "addition" and "$\odot$" of "multiplication"?

15. A set $A = \{x\}$ with one element has just two subsets, and so the ring of all subsets of A, as defined in Example 10, is a ring with two elements. Make addition and multiplication tables for this ring, and compare it with the ring of Example 5.

16. Make addition and multiplication tables for the ring of all subsets of the set $A = \{1, 2\}$. Verify that by a proper choice of notation, this ring is the ring of Example 3.

17. The following is an addition table and part of the multiplication table for a ring with three elements. Make use of the distributive laws to fill in the rest of the multiplication table.

$(+)$	a	b	c
a	a	b	c
b	b	c	a
c	c	a	b

$(\cdot)$	a	b	c
a	a	a	a
b	a	b	—
c	a	—	—

Is this a commutative ring? Does it have a unity?

18. Do the same as in the preceding exercise, using the following addition table and partial multiplication table for a ring with four elements.

$(+)$	a	b	c	d
a	a	b	c	d
b	b	a	d	c
c	c	d	a	b
d	d	c	b	a

$(\cdot)$	a	b	c	d
a	a	a	a	a
b	a	—	—	a
c	a	—	c	—
d	a	b	c	—

Is this a commutative ring? Does it have a unity?

19. If a and b are any integers let us give the following new definitions of "addition" and "multiplication," indicated respectively by "$\oplus$" and "$\odot$":

$$a \oplus b = a + b - 1, \quad a \odot b = a + b - ab.$$

Verify that with respect to these definitions of "addition" and "multiplication," the set of all integers is a commutative ring with unity. What is the zero of this ring?

20. If R and S are rings, give a detailed proof that the direct sum $R \oplus S$ is a ring.

21. (i) If the ring R has m elements and the ring S has n elements (m and n being positive integers), how many elements are there in the direct sum $R \oplus S$?

(ii) Give an example of a commutative ring with 16 elements and an example of a noncommutative ring with 16 elements.

(iii) Give an example of a ring with 32 elements which does not have a unity.

2.4 SOME PROPERTIES OF ADDITION

So far we have given the definition of a ring and have presented a number of examples of rings of many different kinds. It should by now be clear that when we think of an arbitrary ring we should not necessarily think of one of our familiar number systems. Accordingly, we cannot consider any properties of a ring as being obvious, except those actually used in the definition. In this section and the next we shall give proofs of a number of properties of any ring. In the present section, we consider only properties of addition, and hence will use in our proofs only the properties P_1–P_4.

Before proceeding, it may be well to recall that if a and b are elements of a set (in particular, of a ring), by $a = b$ we mean that a and b are identical elements of the set or, looked at another way, a and b are different symbols for the same element. As a consequence of this usage of "equality," it is clear that equality is an equivalence relation. Furthermore, when working in a ring, the definition of a binary operation implies that if $a = b$ and $c = d$, then $a + c = b + d$ and $ac = bd$. In the proofs below, we shall freely use these facts without explicit mention.

First, let us prove the following result.

2.3 Theorem. *The zero of a ring R, whose existence is asserted by P_3, is unique.*

PROOF. By this statement, we mean the following. If $0, 0' \in R$ such that for every element a of R,

$$a + 0 = a \tag{1}$$

and also

$$a + 0' = a, \tag{2}$$

then $0 = 0'$. The proof is as follows. Since Equation (1) is true for every element a of R, we may replace a in this equation by $0'$. Hence, we have that

$$0' + 0 = 0'. \tag{3}$$

In like manner, it follows from Equation (2) that

$$0 + 0' = 0. \tag{4}$$

Since, by the commutative law of addition, $0' + 0 = 0 + 0'$, it follows from Equations (3) and (4) that $0 = 0'$, and the proof is complete.

In view of this result, we are justified in speaking of *the* zero of a ring. An element which is not the zero may naturally be called a *nonzero* element.

We may observe that this is the first theorem which we have proved about an arbitrary ring; that is, the definition of a ring has this result as a logical consequence. Although the truth of this theorem may be easily verified for all the specific examples of rings which have been given, the verification for any number of examples would not in itself constitute a proof that it is always true. Now we know that it must be true in *every* ring. A considerable number of other results for arbitrary rings will be obtained as we proceed.

We next prove the following theorem.

2.4 Theorem (CANCELLATION LAWS OF ADDITION). *If a, b, and c are elements of a ring R, the following are true:*

(*i*) *If $a + c = b + c$, then $a = b$,*
(*ii*) *If $c + a = c + b$, then $a = b$.*

PROOF. We proceed to prove the first statement of this theorem. Let us therefore assume that

$$a + c = b + c. \tag{5}$$

By P_4, there exists an element t of R such that

$$c + t = 0. \tag{6}$$

Now it follows from Equation (5) that

$$(a + c) + t = (b + c) + t. \tag{7}$$

But

$$\begin{aligned} (a + c) + t &= a + (c + t) && (\textit{assoc. law}) \\ &= a + 0 && (\textit{Equation (6)}) \\ &= a && (\textit{definition of 0}). \end{aligned}$$

Similarly,

$$\begin{aligned}(b + c) + t &= b + (c + t)\\ &= b + 0 = b.\end{aligned}$$

From these calculations, and Equation (7), we see that $a = b$, as we wished to show.

In view of the commutative law of addition, part (ii) of the theorem follows at once from part (i).

Although the definition of the zero requires that $a + 0 = a$ for *every* element a of a ring R, we can now observe that the zero of a ring is completely determined by any *one* element. By this statement, we mean that *if d is some one element of R and $d + z = d$, then we must have* $z = 0$. Of course, this fact is an immediate consequence of the preceding theorem. For $d + z = d$ and $d + 0 = d$ imply that $d + z = d + 0$, from which it follows that $z = 0$.

The next result is also an almost immediate consequence of the preceding theorem.

2.5 Corollary. *The additive inverse of an element a of a ring R, whose existence is asserted by Property P_4, is unique.*

PROOF. To prove this statement, suppose that $a + x = 0$ and that $a + y = 0$. Then $a + x = a + y$, and one of the cancellation laws of addition shows at once that $x = y$.

Since each element a of R has exactly one additive inverse, we shall find it convenient to denote this additive inverse by $-a$, and shall also often write $b + (-a)$ in the form $b - a$. It may be helpful to have in mind a verbal definition of $-a$ as follows, "$-a$ is the element of R which when added to a gives 0." That is, if $a + x = 0$ (or, equally well, $x + a = 0$), it follows that $x = -a$.

Since $a + (-a) = 0$, we see also that a is the additive inverse of $-a$, that is, that $-(-a) = a$. We have thus established the first of the following, where a, b, and c are arbitrary elements of a ring:

2.6

(i) $-(-a) = a,$

(ii) $-(a + b) = -a - b,$

(iii) $-(a - b) = -a + b,$

(iv) $(a - b) - c = a - (b + c).$

Let us prove the second of these statements. Now $-(a + b)$ is, by definition, the additive inverse of $a + b$, and we proceed to verify as follows that also $-a - b$ is the additive inverse of $a + b$:

$$\begin{aligned}
(a + b) + (-a - b) &= (a + b) + ((-a) + (-b)) && \textit{(notation)} \\
&= [(a + b) + (-a)] + (-b) && \textit{(assoc. law)} \\
&= [a + (b + (-a))] + (-b) && \textit{(assoc. law)} \\
&= [a + (-a + b)] + (-b) && \textit{(comm. law)} \\
&= [(a + (-a)) + b] + (-b) && \textit{(assoc. law)} \\
&= (0 + b) + (-b) && \textit{(def. of } -a) \\
&= b + (-b) && \textit{(def. of 0)} \\
&= 0 && \textit{(def. of } -b).
\end{aligned}$$

We therefore see that both $-(a + b)$ and $-a - b$ are additive inverses of $a + b$. Hence, the uniqueness of the additive inverse implies that

$$-(a + b) = -a - b,$$

and the proof is complete. The proofs of the other two parts of 2.6 will be given as exercises in the next list of exercises.

The final theorem of this section is the following.

2.7 Theorem. *If a and b are elements of a ring R, the equation $a + x = b$ has in R the* unique *solution $x = b - a$.*

PROOF. It is easy to verify that $x = b - a$ *is* a solution. For

$$\begin{aligned}
a + (b - a) &= a + (-a + b) && \textit{(comm. law)} \\
&= (a + (-a)) + b && \textit{(assoc. law)} \\
&= 0 + b = b.
\end{aligned}$$

The *uniqueness* of the solution follows from one of the cancellation laws. For if we have $a + x = b$ and $a + y = b$, then $a + x = a + y$, and this implies that $x = y$.

2.5 SOME OTHER PROPERTIES OF A RING

In this section we shall establish some properties of a ring that involve multiplication only and some that involve both addition and multiplication.

First, we prove the following result.

2.8 Theorem. *A ring can have at most one unity.*

PROOF. The proof is much like the proof of the uniqueness of the zero. Suppose that $e, e' \in R$ such that for every element a of R,

$$ea = ae = a, \tag{1}$$

and also

$$e'a = ae' = a. \tag{2}$$

In particular, Equation (1) must hold for $a = e'$, that is, we must have

$$ee' = e'e = e'. \tag{3}$$

Similarly, by setting $a = e$ in Equation (2), we obtain

$$e'e = ee' = e. \tag{4}$$

Equations (3) and (4) then imply that $e = e'$, and there is only one unity. If a ring has a unity, we may therefore properly speak of *the* unity of a ring.

We next make the following definition.

2.9 Definition. Let a be an element of a ring R with unity e. If there exists an element s of R such that

$$as = sa = e,$$

then s is called a *multiplicative inverse* of a.

One of the defining properties of a ring states that every element has an additive inverse. However, simple examples show that the situation may be quite different for multiplicative inverses. In the ring of all real numbers it is true that every nonzero element has a multiplicative inverse. In the ring $\mathbf{Z}$ of all integers, there are exactly two elements that have multiplicative inverses, namely, 1 and -1. In the ring of all subsets of a given set A (Example 10), the only element that has a multiplicative inverse is the unity e of the ring, that is, the subset consisting of the entire set A. For if a and b are elements of this ring, ab (which we defined to be $a \cap b$) is a proper subset of A if either a or b is a proper subset of A. Hence, $ab = e$ only if $a = e$ and $b = e$.

In view of these examples, it is clear that we must never take it for granted that an element of a ring necessarily has a multiplicative inverse. However, the following result is easy to establish.

2.10 Theorem. *If an element a of a ring R with unity e has a multiplicative inverse, it is unique.*

PROOF. Suppose that both s and t are multiplicative inverses of the element a. Then, using the fact that $sa = e$ and the associative law of multiplication, we see that

$$s(at) = (sa)t = et = t.$$

But since $at = e$, it is also true that

$$s(at) = se = s,$$

and it follows that $s = t$.

In case a has a multiplicative inverse, it is customary to designate this multiplicative inverse by a^{-1}.

It will be recalled that the zero of a ring has been defined in terms of addition only. However, we shall now prove the following theorem, which has a familiar form.

2.11 Theorem. *For each element a of a ring R, we have*

$$a \cdot 0 = 0 \cdot a = 0.$$

PROOF. Since $a + 0 = a$, it follows that

$$a(a + 0) = a \cdot a.$$

But, by one of the distributive laws,

$$a(a + 0) = a \cdot a + a \cdot 0.$$

Hence,

$$a \cdot a + a \cdot 0 = a \cdot a.$$

Now we know that $a \cdot a + 0 = a \cdot a$ and, by Theorem 2.4, we conclude that $a \cdot 0 = 0$.

In case R is a commutative ring, it follows from what we have just proved that also $0 \cdot a = 0$. If R is not commutative, a proof that $0 \cdot a = 0$ can easily be given using the other one of the distributive laws. This proof will be left as an exercise.

The following can now be verified in turn for arbitrary elements a, b, and c of a ring:

2.12

$$\begin{aligned}
&\text{(i)} & a(-b) &= -(ab), \\
&\text{(ii)} & (-a)b &= -(ab), \\
&\text{(iii)} & (-a)(-b) &= ab, \\
&\text{(iv)} & a(b - c) &= ab - (ac), \\
&\text{(v)} & (b - c)a &= ba - (ca).
\end{aligned}$$

The proof of (i) goes as follows. We have

$$a(b + (-b)) = a \cdot 0 = 0.$$

However, by one of the distributive laws, we know that

$$a(b + (-b)) = ab + a(-b).$$

Hence,

$$ab + a(-b) = 0.$$

But since ab has a unique additive inverse $-(ab)$, it follows that $a(-b) = -(ab)$. The proofs of the other parts of 2.12 will be left as exercises.

In view of 2.12(i) and (ii), we see that

$$-(ab) = (-a)b = a(-b).$$

Accordingly, in later sections we shall usually write simply $-ab$ for any one of these equal expressions.

Let us now make a few remarks about the concept of *subring*. If S is a subring of R, then not only is the set S a subset of the set R but also S must be closed under the operations of addition and multiplication already defined in R. In particular, it follows that the zero of R is also the zero of S and, moreover, the additive inverse of an element of the subring S is identical with the additive inverse of this element considered as an element of R. (Why?) Several of the defining properties of a ring hold in S simply because they hold in R. For example, the associative law of addition clearly holds in S because it holds in R and $S \subseteq R$.

The following theorem, whose proof will be left as an exercise, furnishes a convenient way to determine whether a set of elements of R is actually a subring of R.

2.13 Theorem. *Let R be a ring and S a nonempty subset of the set R. Then S is a subring of R if and only if the following conditions hold:*

(i) S is closed under the operations of addition and multiplication defined on R.

(ii) If $a \in S$, then $-a \in S$.

EXERCISES

1. Prove 2.6(iii) and (iv).
2. If a and b are elements of the ring of all subsets of a given set (Example 10), show that (i) $a = -a$ and (ii) the equation $a + x = b$ has the solution $x = a + b$.
3. Complete the proof of Theorem 2.11 by showing that $0 \cdot a = 0$ for every element a of any ring.
4. Prove 2.12(ii)–(v).
5. If a, b, c, and d are elements of a ring, prove each of the following:

(i) $(a + b)(c + d) = (ac + ad) + (bc + bd)$,
(ii) $(a + b)(c + d) = (ac + bc) + (ad + bd)$,
(iii) $(a - b)(c - d) = (ac + bd) - (bc + ad)$,
(iv) $(a + b)(c - d) = (ac + bc) - (ad + bd)$,
(v) $(a - b)(c + d) = (ac + ad) - (bc + bd)$,
(vi) $(a(-b))(-c) = a(bc)$.

6. Verify that every nonzero element of the ring of Example 7 has a multiplicative inverse.
7. Show that an element (a, b, c) of the ring L of Example 8 has a multiplicative inverse if and only if $a = \pm 1$ and $c = \pm 1$.
8. (i) Find the multiplicative inverse of each of the following elements of the ring W of Example 9:

$$\begin{bmatrix} 2 & 5 \\ 1 & 3 \end{bmatrix}, \quad \begin{bmatrix} 2 & 7 \\ 1 & 3 \end{bmatrix}, \quad \begin{bmatrix} 6 & 1 \\ 17 & 3 \end{bmatrix}.$$

(ii) Show that the element

$$\begin{bmatrix} 1 & 2 \\ 0 & 3 \end{bmatrix}$$

of the same ring W does not have a multiplicative inverse in W.

9. Let R be a ring with unity. If a and b are elements of R that have multiplicative inverses, show that ab has a multiplicative inverse by verifying that $(ab)^{-1} = b^{-1}a^{-1}$.
10. Give an example of elements a and b of some ring such that a^{-1} and b^{-1} exist, but $(ab)^{-1} \neq a^{-1}b^{-1}$.

11. Suppose that a, c, and d are elements of a ring R and that a has a multiplicative inverse in R. Prove that if $ac = ad$ (or $ca = da$), then $c = d$. Show how Theorem 2.10 may be considered to be a special case of this result.

12. Prove Theorem 2.13.

13. If R and S are rings, verify that the set of elements of the direct sum $R \oplus S$ of the form $(a, 0)$, where $\alpha \in R$ and 0 is the zero of S, is a subring of $R \oplus S$.

14. Give an example of a ring R having a subring S such that
- (i) R has a unity and S does not have a unity,
- (ii) R does not have a unity but S has a unity,
- (iii) R and S have the same unity,
- (iv) R and S both have unities, but they are different,
- (v) R is a noncommutative ring and S is commutative.

15. Show that the set of all elements of the ring W of Example 9 of the form

$$\begin{bmatrix} x & 0 \\ y & z \end{bmatrix},$$

where $x, y, z \in \mathbf{Z}$, is a subring of the ring W.

16. If S and T are subrings of a ring R, show that $S \cap T$ is a subring of R. Give an example to show that $S \cup T$ need not be a subring.

17. (i) Give the addition table and multiplication table for a ring with exactly one element.
(ii) If a ring R has more than one element and has a unity e, show that $e \neq 0$.

18. Let R be an arbitrary ring and $a \in R$. Prove that the set $\{x \mid x \in R, ax = 0\}$ is a subring of R.

2.6 GENERAL SUMS AND PRODUCTS

The operations of addition and multiplication are *binary* operations; that is, they apply to *two* elements only. Let us now consider how we can give a meaning to sums or products of three or more elements of a ring.

If a_1, a_2, and a_3 are elements of a ring, let us define $a_1 + a_2 + a_3$ as follows:

2.14 $$a_1 + a_2 + a_3 = (a_1 + a_2) + a_3.$$

However, by the associative law of addition, it then follows that

2.15 $$a_1 + a_2 + a_3 = a_1 + (a_2 + a_3),$$

and therefore a sum of three elements is independent of the way parentheses might be introduced to indicate the manner of association of the elements.

Now that we have defined a sum of three elements of a ring, let us define a sum of four elements as follows:

2.16 $$a_1 + a_2 + a_3 + a_4 = (a_1 + a_2 + a_3) + a_4.$$

The associative law of addition then shows that

2.17 $$a_1 + a_2 + a_3 + a_4 = (a_1 + a_2) + (a_3 + a_4)$$

and also that

2.18 $$a_1 + a_2 + a_3 + a_4 = a_1 + (a_2 + a_3 + a_4).$$

These calculations verify that the sum of four elements is also independent of the way in which the elements may be associated.

It seems fairly clear that similar statements hold for sums of more than four elements. A general proof can be given by the method of mathematical induction. This method of proof will be discussed in the next chapter, at which time we shall return to a further consideration of the material of this section. Although we shall not now give a proof, we proceed to formulate a statement which generalizes what we have said above about sums of three or four elements, as well as some similar statements also involving products.

We have in 2.14 and 2.16 defined the sum of three or four elements of a ring. In general, if k is a positive integer such that

$$a_1 + a_2 + \cdots + a_k$$

has been defined, we define

2.19 $$a_1 + a_2 + \cdots + a_k + a_{k+1} = (a_1 + a_2 + \cdots + a_k) + a_{k+1}.$$

It should then appear that this gives us a definition of the sum of any number n of elements of a ring. Such a definition is called a *recursive* definition, and definitions of this kind will be considered more carefully in the next chapter.

We shall not write out the details, but in precisely the same way it is possible to give a recursive definition of a product of any number n of elements of a ring.

We now state the following theorem, which generalizes several of the properties used in the definition of a ring.

2.20 Theorem. *Let n be an arbitrary positive integer, and let $a_1, a_2, \cdots, a_n$ be elements of a ring R.*

(*i*) Generalized associative laws. *For each positive integer* r *such that* $1 \leq r < n$, *we have*

2.21 $$(a_1 + a_2 + \cdots + a_r) + (a_{r+1} + \cdots + a_n) = a_1 + a_2 + \cdots + a_n$$

and

2.22 $$(a_1a_2 \cdots a_r)(a_{r+1} \cdots a_n) = a_1a_2 \cdots a_n.$$

(*ii*) Generalized distributive laws. *If* $b \in R$, *we have*

2.23 $$b(a_1 + a_2 + \cdots + a_n) = ba_1 + ba_2 + \cdots + ba_n$$

and

2.24 $$(a_1 + a_2 + \cdots + a_n)b = a_1b + a_2b + \cdots + a_nb.$$

(*iii*) Generalized commutative laws. *If* $i_1, i_2, \cdots, i_n$ *are the integers* $1, 2, \cdots, n$ *in any order, then*

2.25 $$a_{i_1} + a_{i_2} + \cdots + a_{i_n} = a_1 + a_2 + \cdots + a_n.$$

In case R is a commutative *ring, we have also*

2.26 $$a_{i_1}a_{i_2} \cdots a_{i_n} = a_1a_2 \cdots a_n.$$

It will be observed that if $n = 3$ in 2.21, then necessarily $r = 1$ or $r = 2$ and the truth of 2.21 in these cases is asserted by 2.15 and 2.14. Similarly, if $n = 4$, it follows that $r = 1, 2$, or 3. These three cases of 2.21 have been verified in 2.18, 2.17, and 2.16.

Let us explain the notation used in the generalized commutative laws. As an example, let $n = 3$, and let $i_1 = 3$, $i_2 = 1$, $i_3 = 2$. Then 2.25 states that

$$a_3 + a_1 + a_2 = a_1 + a_2 + a_3.$$

This special case can readily be verified as follows:

$$\begin{aligned} a_3 + a_1 + a_2 &= (a_3 + a_1) + a_2 && (\textit{def.}) \\ &= a_3 + (a_1 + a_2) && (\textit{assoc. law}) \\ &= (a_1 + a_2) + a_3 && (\textit{comm. law}) \\ &= a_1 + a_2 + a_3 && (\textit{def.}). \end{aligned}$$

As stated above, a general proof of any part of the above theorem requires the use of mathematical induction, and this method of proof will be discussed later. However, the theorem should seem fairly obvious; for the time being we shall merely assume it without proof. In particular, Theorem 2.20(i) assures us that we can introduce parentheses in a sum or product to indicate association in any way we wish without changing the value of the respective sum or product. Accordingly, we shall henceforth usually omit such parentheses entirely. Some special cases of the theorem will be assigned as exercises at the end of this section.

We next observe that positive integral exponents may be defined in any ring R in the usual way. If a is an arbitrary element of R, we may define $a^1 = a$, $a^2 = a \cdot a$, and, in general, if k is a positive integer such that a^k has been defined, we define $a^{k+1} = a^k \cdot a$. The following familiar laws of exponents now hold, where m and n are arbitrary positive integers:

2.27
(i) $$a^m \cdot a^n = a^{m+n},$$
(ii) $$(a^m)^n = a^{mn}.$$

Suppose, now, that $a, b \in R$. Then $(ab)^2 = (ab)(ab)$, and if $ba \neq ab$, $(ab)^2$ may not be equal to a^2b^2. However, if $ba = ab$, it does follow that $(ab)^2 = a(ba)b = a(ab)b = a^2b^2$. In general, it is not difficult to show that if R is a *commutative* ring and m is any positive integer, then

2.28 $$(ab)^m = a^m \cdot b^m.$$

We may remark that *negative* integral exponents can be defined if we restrict attention to elements which have multiplicative inverses. However, we shall postpone any consideration of negative exponents until a later chapter.

We now introduce a convenient notation for *multiples* that parallels the exponent notation for *powers*. If $a \in R$, let us define $1a = a$, $2a = a + a$, and, in general, if k is a positive integer such that ka has been defined, we define $(k + 1)a = ka + a$. If 0 is the zero integer, we define $0a$ to be the zero element of R. Actually, there will be no confusion if the same symbol is used to designate the zero integer and the zero of the ring. Since every element of a ring has an additive inverse, we can easily introduce negative multiples as well as positive multiples. If m is a positive integer, we define $(-m)a$ to be $m(-a)$. Then $(-m)a$ is also seen to be equal to $-(ma)$. Thus, for example,

$$(-2)a = 2(-a) = -a - a = -(a + a) = -(2a).$$

The reader may easily convince himself of the truth of the following, it being understood that m and n are any integers (positive, negative, or zero) and that a and b are arbitrary elements of any ring R:

2.29
(i) $$ma + na = (m + n)a,$$
(ii) $$m(na) = (mn)a,$$
(iii) $$m(a + b) = ma + mb,$$
(iv) $$m(ab) = (ma)b = a(mb),$$
(v) $$(ma)(nb) = (mn)(ab).$$

It should perhaps be emphasized that ma is a convenient way of indicating a certain sum of elements of R. However, since the integer m is not necessarily itself an element of R, it is not correct to think of ma as the product of two elements of R. Hence, for example, 2.29(iii) is not necessarily a consequence of one of the distributive laws.

Again, complete proofs of 2.27, 2.28, and 2.29 require the use of mathematical induction.

We may reiterate our present point of view as follows. Except where explicitly stated, as in a later section where we consider formal proofs of some of the facts mentioned here, we shall henceforth assume the truth of 2.20, 2.27, 2.28, and 2.29. In particular, we shall consider it permissible to write sums or products of more than two elements without use of parentheses.

EXERCISES

1. Verify the truth of 2.21 for the case in which $n = 5$, and therefore $r = 1, 2, 3$, or 4.

2. Verify 2.23 for the case in which $n = 3$.

3. Verify 2.25 for the case in which $n = 4$, $i_1 = 3$, $i_2 = 1$, $i_3 = 4$, $i_4 = 2$.

4. If R is a commutative ring, verify 2.28 for the case in which $m = 3$.

5. If a is any element of the ring of Example 3, verify that $2a = 0$. (The zero of the ring is u.)

6. In the ring W of Example 9, let

$$A = \begin{bmatrix} 1 & 2 \\ 0 & 0 \end{bmatrix} \quad \text{and} \quad B = \begin{bmatrix} 0 & 1 \\ 0 & 1 \end{bmatrix}.$$

Verify that $(AB)^2 \neq A^2B^2$.

7. If x and y are any elements of the ring K of Example 6, verify that $(xy)^2 = x^2y^2$, even though this is not a commutative ring.

8. Show that, by a suitable change of notation, 2.27(i) can be considered to be a special case of 2.22.

9. Show that, by a suitable change of notation, 2.29(i) can be considered to be a special case of 2.21.

10. Show that, by a suitable change of notation, 2.29(iv) can be obtained from 2.23 and 2.24.

11. A ring R is called a *Boolean ring* if $a^2 = a$ for every element a of R. If R is a Boolean ring and $a \in R$, prove that $2a = 0$. Then prove that R is necessarily a commutative ring. [Hint: Consider $(a + b)^2$.]

12. Prove that the direct sum $R \oplus S$ of two rings R and S is a Boolean ring if and only if both R and S are Boolean rings.

13. Give an example of a Boolean ring with 32 elements and an example of a ring with 32 elements which is not a Boolean ring.

14. Let R be an arbitrary ring and consider matrices

$$\begin{bmatrix} a & b \\ c & d \end{bmatrix},$$

where $a, b, c, d \in R$. If addition and multiplication are defined as in Example 9, prove that we obtain a ring. This ring is called the ring of all matrices of order two over the ring R.

15. Prove that if a ring R contains elements s and t such that $st \neq 0$, then the ring of all matrices of order two over R is a noncommutative ring.

16. Let R be a ring and K a subring of R with the special property that if $c \in K$ and $r \in R$, then $cr \in K$ and $rc \in K$. If $a, b \in R$, let us define $a \sim b$ to mean that $a - b \in K$. Prove each of the following:

(i) $\sim$ is an equivalence relation on the set R.

(ii) If a, b, c, and d are elements of R with $a \sim b$ and $c \sim d$, then $(a + c) \sim (b + d)$ and $ac \sim bd$.

17. If there exists an element e_1 of a ring R such that $e_1 a = a$ for every element a of R, e_1 may be called a *left unity* of R. Similarly, e_2 is a *right unity* of R if $ae_2 = a$ for every element a of R. Verify that the set S of all matrices of order two over $\mathbf{Z}$ of the form

$$\begin{bmatrix} a & b \\ 0 & 0 \end{bmatrix}, \qquad a, b \in \mathbf{Z},$$

is a subring of the ring W of all matrices of order two over $\mathbf{Z}$. Then prove each of the following:

(i) The ring S has a left unity but no right unity.

(ii) The ring S has an infinite number of left unities.

18. Using the definitions of the preceding exercise, prove that if a ring R has a *unique* left unity, it is also a right unity (and therefore the unity). [Hint: If e_1 is a left unity and $c \in R$, show that $e_1 + ce_1 - c$ also is a left unity.]

2.7 HOMOMORPHISMS AND ISOMORPHISMS

The concepts to be introduced in this section play an exceedingly important role in modern algebra. Before giving formal definitions, we illustrate the ideas by several examples.

Example 1: Let $\mathbf{Z}$ be the ring of integers and T the ring of Example 5 of Section 2.3, whose addition and multiplication tables we here reproduce for convenience.

$(+)$	0	e
0	0	e
e	e	0

$(\cdot)$	0	e
0	0	0
e	0	e

Now let $\theta: \mathbf{Z} \to T$ be the mapping defined as follows for $i \in \mathbf{Z}$:

$$i\theta = \begin{cases} 0 & \text{if } i \text{ is even,} \\ e & \text{if } i \text{ is odd.} \end{cases}$$

Thus, for example, $4\theta = 0$ and $7\theta = e$. We next observe that

$$(4 + 7)\theta = 11\theta = e = 4\theta + 7\theta,$$

and also that

$$(4 \cdot 7)\theta = 28\theta = 0 = (4\theta)(7\theta).$$

Actually, it may be verified that similar results always hold. That is, if $i, j \in \mathbf{Z}$, then

$$(i + j)\theta = i\theta + j\theta$$

and

$$(ij)\theta = (i\theta)(j\theta).$$

As a matter of fact, the addition and multiplication tables for T are just those which we would get if we thought of 0 as standing for "even" and e for "odd." Thus, for example, $e + e = 0$ and "odd" + "odd" = "even."

What we have observed is that in this example "the image of a sum is the sum of the images." More precisely, if we take the sum of i and j in the ring $\mathbf{Z}$, the image of this sum is the sum in the ring T of the respective images of i and j. This fact is often expressed by saying that the operation of *addition is preserved under the mapping* θ. Similarly, the operation of multiplication is also preserved under this mapping θ.

Now the mapping θ of this example is clearly a mapping of $\mathbf{Z}$ *onto* T, and we have indicated that both the operations of addition and multiplication are preserved under this mapping. According to the definition to be given below, the mapping θ is an example of a *homomorphism* of the ring $\mathbf{Z}$ onto the ring T.

Example 2: Let R and S be arbitrary rings, and let $\phi: R \times S \to R$ be the *projection* of the set $R \times S$ onto R, as defined in Example 2 of Section 1.2. Since the set $R \times S$ becomes a ring $R \oplus S$ under natural definitions of addition and multiplication, ϕ may be considered as a mapping of the ring $R \oplus S$ onto the ring R defined by

$$(r, s)\phi = r, \qquad (r, s) \in R \oplus S.$$

We assert that the operations of addition and multiplication are preserved under the mapping ϕ. That this is true for addition is a consequence of the following simple calculations:

$$\begin{aligned}[(r_1, s_1) + (r_2, s_2)]\phi &= (r_1 + r_2, s_1 + s_2)\phi = r_1 + r_2 \\ &= (r_1, s_1)\phi + (r_2, s_2)\phi.\end{aligned}$$

A similar calculation will verify that multiplication also is preserved under the mapping ϕ. Hence ϕ is a homomorphism of $R \oplus S$ onto R.

Example 3: Let $K = \{a, b, c, d\}$ be the ring of Example 6 of Section 2.5 with addition and multiplication tables which we here reproduce.

$(+)$	a	b	c	d
a	a	b	c	d
b	b	a	d	c
c	c	d	a	b
d	d	c	b	a

$(\cdot)$	a	b	c	d
a	a	a	a	a
b	a	b	c	d
c	a	a	a	a
d	a	b	c	d

Now let $L = \{i, j, k, l\}$, with addition and multiplication on L defined by the following tables.

$(+)$	i	j	k	l
i	k	l	i	j
j	l	k	j	i
k	i	j	k	l
l	j	i	l	k

$(\cdot)$	i	j	k	l
i	i	j	k	l
j	k	k	k	k
k	k	k	k	k
l	i	j	k	l

It can be proved that L is a ring with respect to these definitions of addition and multiplication. At first glance, the rings K and L may seem quite different, but it is not difficult to verify that they are identical except for the notation used. If, in the tables for K we replace a by k, b by i, c by j, and d by l, the tables will coincide except for the order in which the elements are written down. Let us state this fact in a more precise way as follows. Let $\psi: K \to L$ be the mapping defined by

$$a\psi = k, \quad b\psi = i, \quad c\psi = j, \quad d\psi = l.$$

Then not only is ψ a *one-one* mapping of K onto L but also the operations of addition and multiplication are preserved under the mapping ψ. Thus ψ is a homomorphism of K onto L which is a one-one mapping (in contrast to the homomorphisms of the two preceding examples). Such a homomorphism is given the special name of an *isomorphism.*

Let us now give formal definitions of the concepts which have been introduced in the above examples.

2.30 Definition. A mapping $\theta: R \to S$ of a ring R into a ring S is called a *homomorphism* if and only if the operations of addition and multiplication are preserved under θ, that is, if and only if for arbitrary elements a, b, of R, the following hold:

2.31 $$(a + b)\theta = a\theta + b\theta, \quad (ab)\theta = (a\theta)(b\theta).$$

If there exists a homomorphism of R *onto* S, we may say that R is *homomorphic* to S or that S is a *homomorphic image* of R.

The following special case is of such importance that we give it a separate definition.

2.32 Definition. A homomorphism which is a one-one mapping is called an *isomorphism.* If there exists an isomorphism of R *onto* S, we may say that R is *isomorphic* to S or that S is an *isomorphic image* of R.

We may emphasize that, for the sake of generality, we have not required a homomorphism or isomorphism to be an onto mapping. However, whenever we say that R is homomorphic (isomorphic) to S or that S is a homomorphic (isomorphic) image of R, we do mean to imply that it is an onto mapping. Although the more general concept is useful in certain parts of the theory, we shall seldom have occasion to refer to homomorphisms (isomorphisms) that are not onto mappings. The reader should watch carefully to see whether the word "onto" is used.*

It may be worth observing that there always exists a trivial homomorphism of any ring R into any ring S. We have only to define the image of every element of R to be the zero element of S. Of course, this does not assert that S is a homomorphic image of R.

* A homomorphism that is a one-one mapping is also called a *monomorphism.* Likewise, a homomorphism that is an onto mapping is sometimes called an *epimorphism.* Thus, to say that R is isomorphic to S is to say that there exists a mapping $R \to S$ which is both a monomorphism and an epimorphism.

If the mapping $\theta: R \to S$ is an isomorphism of the ring R onto the ring S, it may be verified that the mapping $\theta^{-1}: S \to R$, as defined in Section 1.2, is an isomorphism of S onto R. Accordingly, if R is isomorphic to S, then S is isomorphic to R, and we may sometimes simply say that R and S are *isomorphic rings*. As suggested by the last example above, it should be clear that isomorphic rings may be considered as differing only in the notation used to indicate the elements of the rings. Accordingly, isomorphic rings are sometimes said to be *abstractly identical*.

The most fundamental properties of homomorphisms are stated in the following theorem. It should be kept in mind that an isomorphism is a special case of a homomorphism, so that isomorphisms certainly have the stated properties.

2.33 Theorem. *Let $\theta: R \to S$ be a homomorphism of the ring R into the ring S. Then each of the following is true:*

(*i*) *If 0 is the zero of R, then 0θ is the zero of S.*
(*ii*) *If $a \in R$, then $(-a)\theta = -(a\theta)$.*
(*iii*) *If R has a unity e and θ is an onto mapping, then S has $e\theta$ as unity.*
(*iv*) *Suppose that R has a unity and that θ is an onto mapping. If a is an element of R having a multiplicative inverse, then $(a^{-1})\theta = (a\theta)^{-1}$.*
(*v*) *If R is a commutative ring and θ is an onto mapping, then S is a commutative ring.*

PROOF OF (iii). We need to show that $s(e\theta) = (e\theta)s = s$ for every element s of S. Let s be an arbitrary element of S. Since θ is an onto mapping, there exists at least one element r of R such that $r\theta = s$. Now e is a unity of R, and therefore $re = er = r$. Hence $(re)\theta = (er)\theta = r\theta$. Since multiplication is preserved under the mapping θ, it follows that $(r\theta)(e\theta) = (e\theta)(r\theta) = r\theta$, or $s(e\theta) = (e\theta)s = s$, as required.

We leave the proof of the other parts of the theorem as an exercise.

EXERCISES

In these exercises the examples referred to are those of Section 2.3.

1. Prove Theorem 2.33(i), (ii), (iv), and (v).
2. In the ring K of Example 6, show that $\{a, b\}$ is a subring of K which is isomorphic to the ring of Example 5.

3. If $\theta: R \to S$ is a homomorphism of the ring R into the ring S, prove that the set T of all images of elements of R is a subring of S.
4. Exhibit a homomorphism of the ring of Example 3 onto the ring of Example 5.
5. If R and S are rings, verify that the subring of $R \oplus S$ consisting of all elements of the form $(r, 0)$, $r \in R$, is isomorphic to R.
6. Show that the set of all elements of the ring L of Example 8 of the form $(x, 0, x)$, $x \in \mathbf{Z}$, is a subring of L which is isomorphic to $\mathbf{Z}$.
7. If L is the ring of Example 8, show that the mapping $\theta: L \to \mathbf{Z}$ defined by $(a, b, c)\theta = a$ is a homomorphism of L onto $\mathbf{Z}$.
8. It was shown in a previous exercise that the set of all elements of the ring W of Example 9 of the form

$$\begin{bmatrix} x & 0 \\ y & z \end{bmatrix}, \qquad x, y, z \in Z,$$

is a subring U of W. Show that the mapping $\theta: U \to L$ defined by

$$\begin{bmatrix} x & 0 \\ y & z \end{bmatrix}\theta = (x, y, z), \qquad x, y, z \in \mathbf{Z},$$

is an isomorphism of U onto the ring L of Example 8.
9. Does the ring of Exercise 11 of Section 2.3 have a subring isomorphic to the ring of Example 5?
10. If R is the ring of all subsets of the set $\{1, 2\}$, and S is the ring of all subsets of the set $\{x\}$, exhibit a homomorphism of R onto S.
11. If R and S are rings, prove that the ring $R \oplus S$ is isomorphic to the ring $S \oplus R$.
12. If $\theta: R_1 \to R_2$ and $\phi: S_1 \to S_2$ are, respectively, homomorphisms of R_1 onto R_2 and of S_1 onto S_2, exhibit a homomorphism of $R_1 \oplus S_1$ onto $R_2 \oplus S_2$.
13. Give examples of a ring R without unity and a ring S with unity such that S is a homomorphic image of R.
14. Give examples of a noncommutative ring R and a commutative ring S such that S is a homomorphic image of R.
15. Exhibit a pair of rings, each with four elements, which are not isomorphic. Do the same thing for rings with eight elements.
16. Let R be a ring without unity. On the set $R \times \mathbf{Z}$ let us define addition and multiplication as follows:

$$(a, i) + (b, j) = (a + b, i + j),$$
$$(a, i)(b, j) = (ab + ja + ib, ij).$$

Prove that with respect to this addition and multiplication $R \times \mathbf{Z}$ is a ring with unity and that this ring contains a subring which is isomorphic to R.

17. Let R be the ring of Exercise 19 of Section 2.3. Then as *sets* R and $\mathbf{Z}$ are identical, but they are distinct as *rings* since the operations are different. Show that the mapping $\theta: \mathbf{Z} \to R$ defined by $a\theta = 1 - a$, $a \in \mathbf{Z}$, is an isomorphism of $\mathbf{Z}$ onto R.

18. If $\theta: R \to S$ is a homomorphism of the ring R onto the ring S, prove each of the following:

(i) If A is a subring of R, then the set $A\theta = \{a\theta \mid a \in A\}$ is a subring of S.

(ii) If B is a subring of S, the set C of all elements c of R such that $c\theta \in B$ is a subring of R.

NOTES AND REFERENCES

Simple introductions to the theory of rings will be found in the following books listed in the Bibliography: Burton [24], McCoy [26] and [27]. Most of the books on abstract algebra mentioned in the Bibliography contain at least some treatment of this topic. Additional notes on ring theory will appear at the end of Chapter 11.

Chapter 3

Integral Domains

The properties which we used to define a ring were suggested by simple properties of the integers. However, since we have had numerous examples of commutative rings with unity that bear little resemblance to the ring of integers, it is clear that the system of integers must have some other properties in addition to those which make it a commutative ring with unity. Accordingly, in order to specify in some sense *all* the properties of the ring of integers, we need to consider some properties not mentioned in the previous chapter. In the present chapter we proceed to restrict the rings studied and, eventually, shall have enough properties listed that, in a sense to be described precisely later on, the *only* system which has all these properties is the ring of integers. We may then say that we have obtained a characterization of the ring of integers.

One of the properties that we shall require in characterizing the ring of integers is a property which leads in a natural way to the method of proof by mathematical induction. Accordingly, we shall introduce this important method of proof and use it to establish a few of the results that were stated without proof in Section 2.6.

We shall conclude the chapter with a few remarks about an alternate method of approaching the study of the integers in which all the familiar properties are derived from a few simple properties of the *positive* integers only.

3.1 DEFINITION OF INTEGRAL DOMAIN

We have proved that if 0 is the zero of a ring R, then $a \cdot 0 = 0 \cdot a = 0$ for every element a of R. Of course, this is a familiar property of our elementary number systems. However, in some of the rings previously mentioned there exist elements c and d, both of which are different from zero, such that $cd = 0$. For example, in the ring of Example 3 of Section 2.3 we have $wx = u$, where u is the zero. As another example, consider the ring of all subsets of a given set (Example 10). The empty set is the zero of this ring and, by the definition of multiplication in this ring, if c and d are subsets whose intersection is the empty set, then $cd = 0$. In discussing elements of the type just mentioned, it will be convenient to make the following definition.

3.1 Definition. An element a of a ring R is said to be a *divisor of zero in R* if there exists a *nonzero* element c of R such that $ac = 0$ or a *nonzero* element d of R such that $da = 0$.

It is trivial that the zero of a ring R is a divisor of zero (provided R has more than one element and therefore has a nonzero element to play the role of c or d in the above definition). The elementary number systems have no divisors of zero except the zero or, as we shall say, have no nonzero divisors of zero. An alternate way of stating that a ring R has no nonzero divisors of zero is to say that it has the following property:

$$\text{If } r, s \in R \text{ such that } rs = 0, \text{ then } r = 0 \text{ or } s = 0.$$

We next prove the following simple result.

3.2 Theorem (Cancellation Laws of Multiplication). *If a is not a divisor of zero in a ring R, then each of the following holds:*

(*i*) *If $b, c \in R$ such that $ab = ac$, then $b = c$.*
(*ii*) *If $b, c \in R$ such that $ba = ca$, then $b = c$.*

PROOF. Let us prove part (i) of this theorem. If $ab = ac$, it follows that $a(b - c) = 0$. Then, since a is not a divisor of zero, we must have $b - c = 0$ or $b = c$. Of course, part (ii) follows by a similar argument.

It is important to keep in mind that the cancellation laws of multiplication hold *only if a is not a divisor of zero.*

In most of this chapter we shall be studying rings without nonzero divisors of zero. In such a ring the cancellation laws of multiplication as stated in Theorem 3.2 always hold provided only that $a \neq 0$. Moreover, in order to restrict ourselves for the present to rings more like the ring of integers, we shall also require our rings to be commutative and to have a unity. The next definition gives a convenient way to refer to rings having all of these properties.

3.3 Definition. A ring D with more than one element is called an *integral domain* if it is commutative, has a unity, and has no nonzero divisors of zero.

The most familiar examples of integral domains are the ring of integers, the ring of real numbers, and the ring of complex numbers. The reader may verify that the rings of Examples 1, 2, 5, and 7 of Section 2.3 are integral domains, whereas the rings of Examples 3, 4, 6, and 10 are not integral domains.

3.2 ORDERED INTEGRAL DOMAINS

One important property of the integers that has not been mentioned so far is that they can be *ordered*. If we think of the integers as being exhibited in the following way

$$\cdots, -4, -3, -2, -1, 0, 1, 2, 3, 4, \cdots,$$

and a and b are integers, we say that "a is greater than b" if a occurs to the right of b in the above scheme. It is clear that "a is greater than b" means merely that $a - b$ is a positive integer. This observation suggests that the concept of "order" can be defined in terms of the concept of "positive." We therefore make the following definition.

3.4 Definition. An integral domain D is said to be an *ordered integral domain* if D contains a subset D^+ with the following properties:

(i) If $a, b \in D^+$, then $a + b \in D^+$ (*closed under addition*).
(ii) If $a, b \in D^+$, then $ab \in D^+$ (*closed under multiplication*).
(iii) For each element a of D exactly *one* of the following holds:

$$a = 0, \quad a \in D^+, \quad -a \in D^+ \qquad (\textit{trichotomy law}).$$

The elements of D^+ are called the *positive* elements of D. The nonzero elements of D that are not in D^+ are called the *negative* elements of D.

In view of the definition of an integral domain, the cancellation laws of multiplication (as stated in Theorem 3.2) are always valid in an integral domain so long as $a \neq 0$.

We may emphasize that D^+ is just the notation used to designate a particular subset of an ordered integral domain D. No significance is to be attached to the use of the symbol "+" in this connection.

Obviously, the set $\mathbf{Z}^+$ of positive integers has the properties required of D^+ in the above definition, and hence $\mathbf{Z}$ is an ordered integral domain. However, there are other ordered integral domains such as, for example, the integral domain of all rational numbers or the integral domain of all real numbers. However, not all integral domains are ordered integral domains. For example, we shall prove later on that the integral domain of all complex numbers has no subset with the three properties listed in the preceding definition, and therefore this integral domain is not ordered. See also Exercise 20 at the end of this section.

Now let D be any ordered integral domain, and let D^+ be the set of positive elements of D, that is, the set having the three properties stated in the preceding definition. If $c, d \in D$, we *define* $c > d$ (or $d < c$) to mean that $c - d \in D^+$. Then it is clear that $a > 0$ means that $a \in D^+$, that is, that a is a positive element of D. Similarly, $a < 0$ means that $-a \in D^+$ or that a is a negative element of D. The three properties of Definition 3.4 can then be restated in the following form:

3.5

(i) If $a > 0$ and $b > 0$, then $a + b > 0$.
(ii) If $a > 0$ and $b > 0$, then $ab > 0$.
(iii) If $a \in D$, then exactly one of the following holds:

$$a = 0, \quad a > 0, \quad a < 0.$$

It is now not difficult to verify the following additional properties of inequalities:

3.6

(i) If $a > b$, then $a + c > b + c$ for every $c \in D$.
(ii) If $a > b$ and $c > 0$, then $ac > bc$.
(iii) If $a > b$ and $c < 0$, then $ac < bc$.
(iv) If $a > b$ and $b > c$, then $a > c$.
(v) If $a \neq 0$, then $a^2 > 0$.

The proof of the first of these is as follows. If $a > b$, we have $a - b > 0$. However, $a + c - (b + c) = a - b$ and we see at once that $a + c - (b + c) > 0$, that is, that $a + c > b + c$.

Let us now prove 3.6(v). If $a \neq 0$, then by the form 3.5(iii) of the trichotomy law, either $a > 0$ or $-a > 0$. If $a > 0$, it follows from 3.5(ii) that $a^2 > 0$. If $-a > 0$, the same argument shows that $(-a)^2 > 0$. Since, by 2.12(iii), $(-a)^2 = a^2$, it follows again that $a^2 > 0$.

Proofs of the other parts of 3.6 will be left as exercises.

It is obvious that one can define $a \geq b$ (or $b \leq a$) to mean that either $a = b$ or $a > b$, without specifying which. We shall henceforth use this notation whenever convenient to do so. If $a \geq 0$, it is sometimes convenient to say that a is *nonnegative*. By writing $a < b < c$, we shall mean that $a < b$ and that also $b < c$.

If e is the unity of an ordered integral domain D, we have that $e^2 = e$, and $e^2 > 0$ by Exercise 17(ii) of Section 2.5 and 3.6(v). Thus, *in any ordered integral domain D, the unity is a positive element* (that is, it is an element of D^+).

In any ordered integral domain it is possible to introduce the concept of absolute value in the usual way as follows.

3.7 Definition. Let D be any ordered integral domain and $a \in D$. The *absolute value* of a, written as $|a|$, is defined as follows:

(i) If $a \geq 0$, then $|a| = a$.
(ii) If $a < 0$, then $|a| = -a$.

From this definition it follows that $|0| = 0$ and that if $a \neq 0$, then $|a| > 0$.

EXERCISES

1. If a is a divisor of zero in a commutative ring R, show that ar also is a divisor of zero for every element r of R.
2. Let N be the set of elements of an arbitrary ring R which are *not* divisors of zero. Prove that N is closed under multiplication, and verify by an example that N need not be closed under addition.
3. Prove that if a has a multiplicative inverse a^{-1} in a ring R, then a is not a divisor of zero in R.

4. Verify that each of the following is a divisor of zero in the ring of all matrices of order two over **Z**:

$$\begin{bmatrix} 0 & 1 \\ 0 & 0 \end{bmatrix}, \quad \begin{bmatrix} 1 & 2 \\ 0 & 0 \end{bmatrix}, \quad \begin{bmatrix} 1 & 2 \\ 2 & 4 \end{bmatrix}.$$

5. Determine all divisors of zero in the ring $\mathbf{Z} \oplus \mathbf{Z}$.
6. If R and S are integral domains, prove that their direct sum $R \oplus S$ cannot be an integral domain.
7. Let R be the ring of all subsets of the set $\{x, y, z\}$, and **Z** the ring of integers.
   (i) Determine all those elements of $\mathbf{Z} \oplus R$ which have multiplicative inverses.
   (ii) Determine all those elements of $\mathbf{Z} \oplus R$ which are divisors of zero.
8. Let S be a subring of the ring R. If S has a unity $e \neq 0$ but R does not have a unity, prove that e is a divisor of zero in R.
9. Give at least two examples of rings R and S with the property that there exists a homomorphism θ of R onto S and an element a of R which is a divisor of zero in R but $a\theta$ is not a divisor of zero in S.
10. If a commutative ring R has a subset R^+ with the properties specified for D^+ in Definition 3.4, we may naturally call R an *ordered ring*.
   (i) Give an example of an ordered ring which is not an integral domain.
   (ii) Prove that if R has a nonzero divisor of zero, R cannot be an ordered ring.
11. Prove 3.6(ii), (iii), (iv).

In Exercises 6–12, the letters a, b, c, and d represent elements of an ordered integral domain.

12. Prove that if $a > b$, then $-a < -b$.
13. Prove that if $a > b$ and $c > d$, then $a + c > b + d$.
14. Prove that if a, b, c, and d are all positive with $a > b$ and $c > d$, then $ac > bd$.
15. Prove that if $a > 0$ and $ab > ac$, then $b > c$.
16. Prove that $|ab| = |a| \cdot |b|$.
17. Prove that $-|a| \leq a \leq |a|$.
18. Prove that $|a + b| \leq |a| + |b|$.
19. Prove: There cannot be a greatest element in an ordered integral domain D (that is, for each $d \in D$ there exists $c \in D$ such that $c > d$).
20. Use the result of the preceding exercise to give a convincing argument (a formal proof is not required) why an integral domain with a finite number of elements cannot be an ordered integral domain.
21. If Property 3.2(i) holds for every nonzero element a of a ring R, prove that R has no nonzero divisor of zero.

22. If Property 3.2(i) holds for every nonzero element a of a ring R, prove that Property 3.2(ii) also holds for every nonzero element a of R.

23. Prove that in a Boolean ring, as defined in Exercise 11 of Section 2.6, every nonzero element except the unity (if it has a unity) is a divisor of zero.

24. Prove that an isomorphic image of an integral domain is an integral domain.

3.3 WELL-ORDERING AND MATHEMATICAL INDUCTION

We need one further condition to characterize the ring of integers among the ordered integral domains. Let us first make the following general definition.

3.8 Definition. A set S of elements of an ordered integral domain is said to be *well-ordered* if each nonempty subset U of S contains a least element, that is, if for each nonempty subset U of S there exists an element a of U such that $a \leq x$ for every element x of U.

It is apparent that the set of all positive integers is well-ordered, and we shall presently find that this property is precisely what distinguishes the ring of integers from other ordered integral domains. The rational numbers will be considered in detail later on in this book, but we may observe now that the set of positive rationals is not well-ordered. In fact, the set of all positive rational numbers has no least element. For if r is any positive rational number, then $r/2$ is also a positive rational number and $r/2 < r$. Hence there can be no least positive rational number.

Let us pause to clarify our point of view about the ring of integers. We have not *proved* any property of the integers; instead we have from time to time merely assumed that they have certain properties. We are now able to state precisely as follows just what properties of the integers we do wish to consider as known. *We assume that the ring of integers is an ordered integral domain in which the set of positive elements is well-ordered.* Accordingly, when we shall henceforth speak of a proof of any property of the integers we shall mean a proof based on this assumption only. The theorem of the next section will indicate why no other properties are required.

We pointed out in Section 3.2 that if e is the unity of an ordered integral domain D, then $e \in D^+$. If D^+ is well-ordered, we now assert that *e is the least element of D^+*. To see this, suppose that c is the least element of D^+ and that $0 < c < e$. It follows by 3.6(ii) that $0 < c^2 < c$, since $ce = c$. Thus we would have $c^2 \in D^+$, with $c^2 < c$. But this violates the assumption that c is the least elements of D^+, and we have the desired conclusion. In particular, our assumption about the ring of integers assures us that 1 is the least positive integer.

The following theorem, which is the basis of proofs by mathematical induction, is just as "obvious" as the fact that the set of positive integers is well-ordered. However, in accordance with our chosen point of view, we shall give a proof of this result.

3.9 Theorem. *Let K be a set of positive integers with the following two properties:*

(*i*) $1 \in K$.
(*ii*) *If k is any positive integer such that $k \in K$, then also $k + 1 \in K$.*

Then K consists of the set of all *positive integers.*

PROOF. To prove this theorem, let us assume that there is a positive integer not in K, and obtain a contradiction. Let U be the set of all positive integers not in K and therefore, by our assumption, U is not empty. Then, by the well-ordering property, U must contain a least element m. Since, by (i), we have $1 \in K$, clearly $m \neq 1$ and it follows that $m > 1$ and therefore $m - 1 > 0$. Moreover, $m - 1 \in K$ since m was chosen to be the least element of U. Now, by (ii) with $k = m - 1$, we see that $m \in K$. But $m \in U$, and we have obtained the desired contradiction. The proof is therefore complete.

The most frequent application of Theorem 3.9 is to a proof of the following kind. Suppose that there is associated with each positive integer n a *statement* (or proposition) S_n, which is either true or false, and suppose we wish to prove that the statement S_n is true for every positive integer n. Let K be the set of all positive integers n such that S_n is a true statement. If we can show that $1 \in K$, and that whenever $k \in K$ then also $k + 1 \in K$, it will follow from Theorem 3.9 that K is the set of all positive integers. Since $n \in K$ merely means that S_n is true, we may reformulate these remarks in the following convenient form.

3.10 Induction Principle. *Suppose that there is associated with each positive integer n a statement S_n. Then S_n is true for every positive integer n provided the following hold:*

(*i*) *S_1 is true.*

(*ii*) *If k is any positive integer such that S_k is true, then also S_{k+1} is true.*

A proof making use of the Induction Principle (or of Theorem 3.9) is usually called a proof by induction or a proof by mathematical induction.

We have proved 3.9 (and 3.10) on the assumption that the set of positive integers is well-ordered. Exercise 10 below will show that, in fact, the Induction Principle is equivalent to the requirement that the set of positive integers be well-ordered.

We may remark that there is another useful form of the Induction Principle in which condition (ii) is replaced by a somewhat different condition. (See Exercise 9 at the end of this section.)

As a first illustration of the language and notation just introduced, we consider a simple example from elementary algebra. If n is a positive integer, let S_n be the statement that

$$2 + 4 + 6 + \cdots + 2n = n(n + 1),$$

it being understood that the left side is the sum of the first n positive even integers. We now prove that S_n is true for every positive integer n, by verifying (i) and (ii) of 3.10. Clearly, S_1 is true since S_1 merely states that $2 = 1 \cdot 2$. Suppose, now, that k is any positive integer such that S_k is true, that is, such that the following is true:

$$2 + 4 + 6 + \cdots + 2k = k(k + 1).$$

Then, by adding the next even integer, $2(k + 1)$, to both sides we obtain

$$\begin{aligned} 2 + 4 + 6 + \cdots + 2k + 2(k + 1) &= k(k + 1) + 2(k + 1) \\ &= (k + 1)(k + 2). \end{aligned}$$

However, this calculation shows that S_{k+1} is true, and hence we have verified both (i) and (ii) of 3.10. The Induction Principle then assures us that S_n is true for every positive integer n.

We now consider again part of the material of Section 2.6, and we first illustrate by a simple explanation how a recursive definition really involves the Induction Principle. The recursive definition of a^n, which was given earlier, may be stated in the following formal way.

3.11 Definition. If a is an element of a ring R, we define $a^1 = a$. Moreover, if k is a positive integer such that a^k is defined, we define $a^{k+1} = a^k \cdot a$.

Now let S_n be the statement, "a^n is defined by 3.11." The Induction Principle then shows that S_n is true for every positive integer n, that is, that a^n is defined by 3.11 for every positive integer n.

Let us now prove (2.27(i)) that if m and n are arbitrary positive integers, then

3.12 $$a^m \cdot a^n = a^{m+n}.$$

Let S_n be the statement that for the positive integer n, 3.12 is true for *every* positive integer m. Then, by definition of a^{m+1}, we see that $a^m \cdot a^1 = a^{m+1}$, and hence S_1 is true. Let us now assume that k is a positive integer such that S_k is true, that is, such that

3.13 $$a^m \cdot a^k = a^{m+k}$$

for every positive integer m. Then

$$\begin{aligned} a^m \cdot a^{k+1} &= a^m \cdot a^k \cdot a && (\textit{by def. of } a^{k+1}) \\ &= a^{m+k} \cdot a && (\textit{by 3.13}) \\ &= a^{m+k+1} && (\textit{by def. of } a^{(m+k)+1}). \end{aligned}$$

We have now shown that S_{k+1} is true, and the Induction Principle then assures us that S_n is true for every positive integer n. Thus we have given a formal proof of the very familiar law of exponents stated in 3.12. In the above proof we have tacitly made use of the associative law of multiplication. As a matter of fact, it was pointed out in the preceding chapter that 3.12 is actually a special case of the generalized associative law of multiplication.

As a further illustration of the use of mathematical induction in proving the results stated in Section 2.6, we shall prove the generalized associative law of addition (2.21). For convenience of reference, let us first restate in a slightly different notation the recursive definition (2.19) of a sum of more than two elements of a ring. If l is a positive integer and $b_1, b_2, \cdots, b_{l+1}$ are elements of a ring such that

$$b_1 + b_2 + \cdots + b_l$$

is defined, we define

3.14 $$b_1 + b_2 + \cdots + b_{l+1} = (b_1 + b_2 + \cdots + b_l) + b_{l+1}.$$

Now let S_n be the statement that for arbitrary elements a_1, $a_2, \cdots, a_n$ of a ring and for each positive integer r such that $1 \leq r < n$, we have

3.15 $$(a_1 + \cdots + a_r) + (a_{r+1} + \cdots + a_n) = a_1 + a_2 + \cdots + a_n.$$

To establish the generalized associative law of addition, we need to prove that S_n is true for every positive integer n. Clearly, S_1 and S_2 are true, and we verified S_3 and S_4 in Section 2.6. We complete the proof by showing that if k is a positive integer such that S_k is true, then also S_{k+1} is true. Otherwise expressed, if S_k is true and r is an integer such that $1 \leq r < k + 1$, we shall know that

3.16 $$(a_1 + \cdots + a_r) + (a_{r+1} + \cdots + a_{k+1}) \\ = a_1 + a_2 + \cdots + a_{k+1}.$$

The case in which $r = k$ is true at once by definition (3.14) of the right side of 3.16. Suppose, then, that $r < k$. As a special case of 3.14, we have

$$a_{r+1} + \cdots + a_{k+1} = (a_{r+1} + \cdots + a_k) + a_{k+1}.$$

This is used in the first step of the following calculation:

$$\begin{aligned} &(a_1 + \cdots + a_r) + (a_{r+1} + \cdots + a_{k+1}) \\ &\quad = (a_1 + \cdots + a_r) + ((a_{r+1} + \cdots + a_k) + a_{k+1}) \\ &\quad = ((a_1 + \cdots + a_r) + (a_{r+1} + \cdots + a_k)) + a_{k+1} && \textit{(by assoc. law)} \\ &\quad = (a_1 + \cdots + a_k) + a_{k+1} && \textit{(by } S_k\textit{)} \\ &\quad = a_1 + \cdots + a_{k+1} && \textit{(by 3.14)}. \end{aligned}$$

This calculation establishes 3.16 and completes the proof.

In a similar manner the other results that were stated in Section 2.6 can be established by induction. Some of them are listed in the following set of exercises.

EXERCISES

1. Prove the generalized distributive law (2.23):

$$b(a_1 + a_2 + \cdots + a_n) = ba_1 + ba_2 + \cdots + ba_n.$$

2. Prove (2.27(ii)) that for arbitrary positive integers m and n,

$$(a^m)^n = a^{mn}.$$

3. If a and b are elements of a commutative ring, prove (2.28) that $(ab)^m = a^m b^m$ for every positive integer m.

4. If n is a positive integer and $a_1, a_2, \cdots, a_n$ are elements of an integral domain such that $a_1a_2 \cdots a_n = 0$, show that at least one of the a's is zero.
5. If $\theta: R \to S$ is a homomorphism of the ring R into the ring S and $a \in R$, prove that $a^n\theta = (a\theta)^n$ for every positive integer n.
6. Prove (2.29(iii)) that if a and b are elements of a ring, for every integer m (positive, negative, or zero),

$$m(a + b) = ma + mb.$$

7. Prove (2.29(iv)) that if a and b are elements of a ring, for every integer m,

$$m(ab) = (ma)b = a(mb).$$

8. Prove (2.29(i)) that if a is an element of a ring, for all integers m and n,

$$ma + na = (m + n)a.$$

[Hint: Make a number of cases as follows: either m or n is zero; both m and n are positive; one of m, n is positive and the other negative; both m and n are negative.]
9. Use the fact that the set of positive integers is well-ordered to prove the following alternate form of the Induction Principle:

Suppose that there is associated with each positive integer n a statement S_n. Then S_n is true for every positive integer n provided the following hold:

(i) S_1 is true.

(ii) If k is a positive integer such that S_i is true for every positive integer $i < k$, then also S_k is true.
10. *Assume* the truth of the Induction Principle (3.10), and prove that the set of positive integers must be well-ordered. [Hint: Let S_n be the statement, "If a set of positive integers contains an element $\leq n$, then the set contains a least element."]

3.4 A CHARACTERIZATION OF THE RING OF INTEGERS

The purpose of this section is to prove the following theorem.

3.17 Theorem. *Let both D and D' be ordered integral domains in which the set of positive elements is well-ordered. Then D and D' are isomorphic.*

Since we are assuming that **Z** is an ordered integral domain in which the set of positive elements is well-ordered, this theorem will show that $\mathbf{Z}$ is the *only* ring with these properties (if we do not consider isomorphic rings as "different" rings).

As a first step in the proof, we shall prove a lemma. In the statement of this lemma, and henceforth whenever it is convenient to do so, we shall make use of the notation introduced in Definition 3.4 and let $\mathbf{Z}^+$ denote the set of all positive integers.

3.18 Lemma. *Let D be an ordered integral domain in which the set D^+ of positive elements is well-ordered. If e is the unity of D, then*

$$D^+ = \{me \mid m \in \mathbf{Z}^+\}$$

and

$$D = \{ne \mid n \in \mathbf{Z}\}.$$

Moreover, if $n_1, n_2 \in \mathbf{Z}$ such that $n_1e = n_2e$, then $n_1 = n_2$.

PROOF. We recall that $a \in D^+$ can also be expressed by writing $a > 0$. We have already observed that $e > 0$ and, in fact, that e is the least element of D^+. For each positive integer n, let S_n be the statement that $ne > 0$. Since, by definition, $1e = e$, we see that S_1 is true. If, now, k is a positive integer such that S_k is true, it follows from 3.5(i) that $(k + 1)e = ke + e > 0$, and therefore that S_{k+1} is true. We have therefore proved by mathematical induction that $me > 0$ for every positive integer m. That is, $me \in D^+$ for every positive integer m. We now proceed to show that all elements of D^+ are of this form. Suppose that this is false and let U be the nonempty set of elements of D^+ that are *not* of this form. Then, since D^+ is well-ordered, U has a least element, say d. Since $e > 0$, and therefore $d - (d - e) > 0$, it follows that $d - e < d$ and hence that $d - e \notin U$; therefore $d - e = m_1e$ for some positive integer m_1. It then follows that $d = e + m_1e = (1 + m_1)e$, and $1 + m_1$ is a positive integer. But d, being an element of U, is *not* of this form, and we have a contradiction. It follows that U must be the empty set, that is, that every element of D^+ is of the required form.

It is now easy to complete the proof of the first statement of the lemma. If $a \in D$, and $a \notin D^+$, then 3.4(iii) implies that $a = 0$ or $-a \in D^+$. If $a = 0$, then $a = 0 \cdot e$. If $-a \in D^+$, then by what we have just proved, $-a = m_2e$ for some positive integer m_2. It follows that $a = (-m_2)e$, and so every element of D is of the form ne, where n is an integer (positive, negative, or zero).

Now suppose that $n_1, n_2 \in \mathbf{Z}$ such that $n_1e = n_2e$. If $n_1 \neq n_2$, we can assume that the notation is so chosen that $n_1 > n_2$. It follows that $n_1 - n_2 > 0$ and, by the part of the lemma already proved, $(n_1 - n_2)e \in D^+$. Hence $(n_1 - n_2)e \neq 0$, or $n_1e \neq n_2e$. Thus the assumption that $n_1 \neq n_2$ leads to a contradiction, and we conclude that $n_1 = n_2$. This completes the proof of the lemma.

It is now easy to prove Theorem 3.17. If e and e' are the respective unities of D and of D', the lemma shows that

$$D = \{ne \mid n \in \mathbf{Z}\}$$

and

$$D' = \{ne' \mid n \in \mathbf{Z}\}.$$

Moreover, the last statement of the lemma asserts that the elements of D are *uniquely* expressible in the form ne, $n \in \mathbf{Z}$. Of course, the elements of D' are likewise uniquely expressible in the form ne', $n \in \mathbf{Z}$.

We now assert that the mapping $\theta: D \to D'$, defined by

$$(ne)\theta = ne', \qquad n \in \mathbf{Z},$$

is the desired isomorphism of D onto D'. By the uniqueness property just obtained, θ is a one-one mapping of D onto D'. Moreover, under this mapping we have

$$\begin{aligned}(n_1e + n_2e)\theta = [(n_1 + n_2)e]\theta &= (n_1 + n_2)e' \\ &= n_1e' + n_2e' = (n_1e)\theta + (n_2e)\theta\end{aligned}$$

and

$$\begin{aligned}[(n_1e)(n_2e)]\theta = [(n_1n_2)e]\theta &= (n_1n_2)e' \\ &= (n_1e')(n_2e') = [(n_1e)\theta][(n_2e)\theta].\end{aligned}$$

Hence, addition and multiplication are preserved and we indeed have an isomorphism. This completes the proof of the theorem.

EXERCISE

1. Let $D = 2^n$, $n \in \mathbf{Z}$ and let us define $2^n \oplus 2^m = 2^{n+m}$ and $2^n \odot 2^m = 2^{nm}$. Prove that with respect to the operations $\oplus$ and $\odot$ so defined, D is an ordered integral domain and that, in fact, D is isomorphic to $\mathbf{Z}$.

★ 3.5 THE PEANO AXIOMS

So far, we have merely *assumed* that the system of all integers has the properties of an ordered integral domain in which the set of positive elements is well-ordered. In this section we shall briefly indicate how it is possible to assume as a starting point only a few simple properties of the natural numbers (positive integers) and then to *prove* all the other properties that are required. This program was first carried out by the Italian mathematician G. Peano, and the simple properties with which we start are therefore called Peano's Axioms. If we denote by N the set of all natural numbers, these axioms are often stated as follows.*

AXIOM 1. $1 \in N$.

AXIOM 2. To each element m of N there corresponds a unique element m' of N called the *successor* of m.

AXIOM 3. For each $m \in N$ we have $m' \neq 1$. (That is, 1 is not the successor of any natural number.)

AXIOM 4. If $m, n \in N$ such that $m' = n'$, then $m = n$.

AXIOM 5. Let K be a set of elements of N. Then $K = N$ provided the following two conditions are satisfied:

(i) $1 \in K$.
(ii) If $k \in K$, then $k' \in K$.

This last axiom is essentially our Theorem 3.9 and is the basis of proofs by mathematical induction. In this approach to the study of the natural numbers it is taken as one of the defining properties or axioms.

Using only these five simple axioms, it is possible to *define* addition and multiplication on N and then to *prove* that N has all the properties of an integral domain except that it does not have a zero and its elements do not have additive inverses. We proceed to give the definitions of addition and multiplication, but shall not carry out the rest of the program. The details can be found in many algebra texts.

The definition of addition is as follows.

* In more modern language, Axioms 2, 3, and 4 simply assert that there exists a one-one mapping of N into N with the property that the element 1 does not occur as an image. Using Axiom 5, it is easy to show that every other element of N is an image.

3.19 Definition. Let m be an arbitrary element of N. First, we define $m + 1 = m'$. Moreover, if $k \in N$ such that $m + k$ is defined, we define $m + k' = (m + k)'$.

By Axiom 5, it follows that the set of all elements n of N such that $m + n$ is defined by 3.19 is the set of *all* elements of N. In other words, an operation of addition is now defined on N.

The operation of multiplication is defined in a similar way as follows.

3.20 Definition. Let m be an arbitrary element of N. First, we define $m \cdot 1 = m$. Moreover, if $k \in N$ such that $m \cdot k$ is defined, we define $m \cdot k' = m \cdot k + m$.

It is also possible to define an order relation on N as follows. If $m, n \in N$, we define $m > n$ to mean that there exists an element k of N such that $m = n + k$. It can then be proved that "$>$" has all the properties that we would expect it to have when applied to the positive elements of an ordered integral domain. Moreover, the set N is well-ordered according to our Definition 3.8 of this concept.

Up to this point we have outlined a program for using Peano's Axioms to establish all the familiar properties of the natural numbers or positive integers. In order to obtain the *ring* of all integers we still have to introduce into the system the negative integers and zero. This can be done by a method quite similar to that which we shall use in Chapter 5 to construct the rational numbers from the integers. Accordingly, we postpone any further discussion of this program until Section 5.7 at the end of Chapter 5.

NOTES AND REFERENCES

Section 5.7 of this book contains a brief discussion as to how one can proceed to obtain the system of all integers from the system of natural numbers. Detailed discussions of the development of the integers from a set of axioms (that of Peano or some other set) will be found, e.g., in Dubish [4], Johnson [7] (appendix), or in the classic work of E. Landau, *Foundations of Analysis*, Chelsea, New York, 1960.

Chapter 4

Some Properties of the Integers

Now that we have obtained a characterization of the ring $\mathbf{Z}$ of integers, in this chapter we proceed to establish a number of simple properties of this system. In giving illustrative examples we shall naturally make use of our familiar decimal notation, but the proofs will be based only on the fact that the ring of integers is an ordered integral domain in which the set of positive elements is well-ordered. In particular, mathematical induction will play a central role in many of the proofs, although we shall frequently omit some of the details.

4.1 DIVISORS AND THE DIVISION ALGORITHM

We begin with the following familiar definition.

4.1 Definition. If $a, d \in \mathbf{Z}$ with $d \neq 0$, d is said to be a *divisor* (or *factor*) of a if there exists an element a_1 of $\mathbf{Z}$ such that $a = a_1d$. If d is a divisor of a, we say also that *d divides a* or that a is *divisible by d* or that a is a *multiple* of d.

We could just as well allow d to be zero in the above definition, but this case is unimportant and it is convenient to exclude it.

We shall often write $d|a$ to indicate that d divides a. We now list as follows a number of simple facts involving the concept of divisor.

(*i*) *If* $d|a$ *and* $a|b$, *then* $d|b$.
(*ii*) $d|a$ *if and only if* $d|(-a)$.
(*iii*) $d|a$ *if and only if* $(-d)|a$.
(*iv*) $\pm 1|a$ *for every integer* a.
(*v*) $d|0$ *for every nonzero integer* d.
4.2 (*vi*) *If* $a \neq 0$ *and* $d|a$, *then* $|d| \leq |a|$. *Moreover, if* $a \neq 0$, $d|a$, *and* $d \neq \pm a$, *then* $|d| < |a|$.
(*vii*) *If* $d|\pm 1$, *then* $d = \pm 1$.
(*viii*) *If* $a|b$ *and* $b|a$, *then* $a = \pm b$.
(*ix*) *If* $d|a$ *and* $d|b$, *then* $d|(ax + by)$ *for arbitrary integers* x *and* y.

We shall prove a few of these statements and leave the proof of the others as an exercise.

PROOF OF 4.2(vi). In the preceding chapter we have proved the not surprising fact that the unity 1 of $\mathbf{Z}$ is the smallest positive integer. Now if $d|a$ there exists an integer a_1 such that $a = a_1 d$, and $a \neq 0$ implies that $a_1 \neq 0$ and $d \neq 0$. Hence $|a_1| \geq 1$ and $|a_1| \cdot |d| \geq |d|$. However, using the well-known result of Exercise 10 of Section 3.2, it follows that $|a| = |a_1 d| = |a_1| \cdot |d| \geq |d|$. This completes the proof of the first statement of 4.2(vi). The second statement follows from the part just proved and the observation that $|d| = |a|$ if and only if $d = \pm a$.

PROOF OF 4.2(vii). If $d|\pm 1$, then 4.2(vi) shows that $|d| \leq |\pm 1| = 1$. Since $d \neq 0$, we must have $|d| = 1$, or $d = \pm 1$.

PROOF OF 4.2(ix). Since $d|a$ and $d|b$, there exist integers a_1 and b_1 such that $a = a_1 d$ and $b = b_1 d$. Hence, if $x, y \in \mathbf{Z}$, we have $ax + by = (a_1 x + b_1 y)d$ and therefore $d|(ax + by)$.

The following concept is an important one in studying divisibility properties of the integers.

4.3 Definition. A nonzero integer p other than 1 or -1 is called a *prime* if its only divisors are ± 1 and $\pm p$.

If $n > 1$, it follows from this definition that n is *not* a prime if and only if there exist positive integers n_1 and n_2, with $1 < n_1 < n$ and $1 < n_2 < n$, such that $n = n_1 n_2$.

It is obvious that $-p$ is a prime if and only if p is a prime. The first few positive primes are

$$2, 3, 5, 7, 11, 13, 17, 19, \cdots.$$

One of the principal reasons for the importance of the primes is that every integer other than 0, 1, and -1 is a prime or can be expressed as a product of primes. This is sometimes taken for granted in arithmetic, and no doubt seems almost obvious. However, the fact that any integer can be so expressed and in only one way, in a sense to be made precise later, is not trivial to prove and is so important that it is often called the "Fundamental Theorem of Arithmetic." We shall return to this theorem in a later section of this chapter.

We proceed to a consideration of the following result.

4.4 Division Algorithm. *If $a, b \in \mathbf{Z}$ with $b > 0$, there exist unique integers q and r such that*

4.5 $$a = qb + r, \qquad 0 \leq r < b.$$

Before giving a detailed proof, we can make the existence of q and r appear plausible by use of a geometric argument. Consider a coordinate line with the multiples of b marked off as in Figure 8. Then

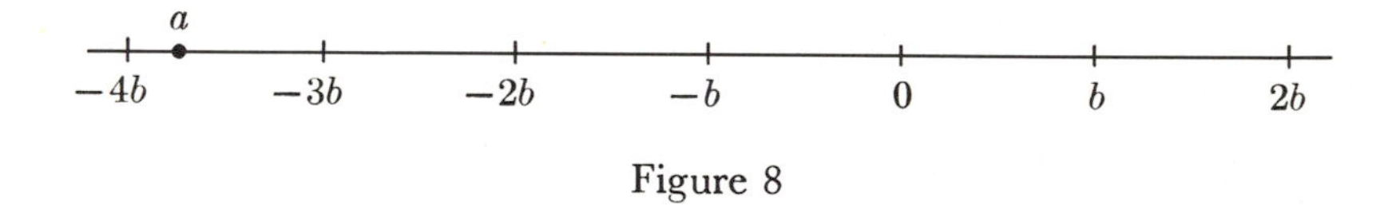

Figure 8

if a is marked off, it either falls on a multiple of b, say qb, or it falls between two successive multiples of b, say qb and $(q + 1)b$. (In the figure, $q = -4$.) In either case, there exists an integer q such that $qb \leq a < (q + 1)b$. If we set $r = a - qb$, then $a = qb + r$ and it is clear that $0 \leq r < b$.

PROOF. Let us now give a proof which does not make use of geometric intuition. An outline of the proof is as follows. Since 4.5 can be written in the form $r = a - qb$, we consider those nonnegative integers of the form $a - xb$, where $x \in \mathbf{Z}$, and shall show that one of them is necessarily less than b. This one we shall then identify with the integer r, whose existence we wish to establish. In order to carry out the details, let S be the set of integers defined as follows:

$$S = \{a - xb \mid x \in \mathbf{Z}, a - xb \geq 0\}.$$

First, we show that the set S is not empty. Now $b \geq 1$, since b is assumed to be a positive integer. It follows that $|a| \cdot b \geq |a|$, and hence that $a + |a| \cdot b \geq a + |a| \geq 0$. Hence, by using $x = -|a|$, we see that S contains the integer $a + |a| \cdot b$, and is therefore not empty. If $0 \in S$, clearly 0 is the least element of S. If $0 \notin S$, S is a nonempty set of positive integers and therefore has a least element, since the set of positive integers is well-ordered. Hence, in either case, S has a least element, say r. There must then exist an integer q such that $a - qb = r$. We therefore have $a = qb + r$, where $0 \leq r$, and we proceed to show that also $r < b$. Let us suppose, on the contrary, that $r \geq b$. Then $r - b \geq 0$ and, since $r - b = a - (q + 1)b$, we see that $r - b \in S$. But since $b > 0$, $r - b < r$ and we have a contradiction of the fact that r is the least element of S. Hence $r < b$, and this completes the proof that there exist integers q and r satisfying 4.5. It remains to be proved that they are unique.

Suppose that q and r satisfy 4.5, and that also

$$a = q_1 b + r_1, \qquad 0 \leq r_1 < b.$$

Then $qb + r = q_1 b + r_1$, and it follows that

4.6 $$b(q - q_1) = r_1 - r.$$

Using the fact that the absolute value of a product is equal to the product of the absolute values, and that $|b| = b$, we obtain

$$b \cdot |q - q_1| = |r_1 - r|.$$

But, since $0 \leq r < b$ and $0 \leq r_1 < b$, we must have $|r_1 - r| < b$ and it follows that $b \cdot |q - q_1| < b$. Since $|q - q_1|$ is a nonnegative integer, this implies that $|q - q_1| = 0$, that is, that $q = q_1$; and 4.6 then shows that also $r = r_1$. We have therefore proved the uniqueness of the integers q and r satisfying 4.5.

The unique integers q and r which satisfy 4.5 are called, respectively, the *quotient* and the *remainder* in the division of a by b. It is important to observe that a is divisible by b if and only if the remainder in the division of a by b is zero.

In a numerical case, at least if $a > 0$, the actual calculation of q and r can be carried out by the familiar process of long division. The method is easily adapted also to the case in which a is negative.

EXERCISES

1. Prove 4.2(i)–(v), (viii).
2. Show that if $x = y + z$ and d is a divisor of any two of the integers x, y, and z, it is also a divisor of the third.
3. Let b and m be positive integers. If q is the quotient and r is the remainder when the integer a is divided by b, show that q is the quotient and mr is the remainder when ma is divided by mb.
4. If p and q are positive primes such that $p|q$, show that $p = q$.
5. If n is a positive integer and $p_1, p_2, \cdots, p_n$ are distinct positive primes, show that the integer $(p_1p_2 \cdots p_n) + 1$ is divisible by none of these primes.
6. For each of the following pairs of integers find the quotient and the remainder in the division of the first integer by the second, and verify Equation 4.5:
   (i) 1251, 78 (ii) 31, 158
   (iii) 4357, 418 (iv) -168, 15.
7. Prove the following generalized form of the Division Algorithm. If $a, b \in \mathbf{Z}$ with $b \neq 0$, there exist unique integers q and r such that $a = qb + r$, $0 \leq r < |b|$. [Hint: Make use of the case already proved in which b was assumed to be positive.]
8. Give a formal proof that if $i \in \mathbf{Z}$, the largest integer which is less than i is $i - 1$, that is, that there exists no integer j such that $i - 1 < j < i$.
9. Let a, b, and c be integers with $b > 0$ and $c > 0$. If q is the quotient when a is divided by b and q' is the quotient when q is divided by c, prove that q' is the quotient when a is divided by bc.

★ 4.2 DIFFERENT BASES

In this section we make use of the Division Algorithm to prove a result which, although independent of the rest of this book, is of some interest in itself.

It is customary to use the integer 10 as the base of our number system. By this we mean that when we write 4371, for example, it is understood to stand for $4 \cdot 10^3 + 3 \cdot 10^2 + 7 \cdot 10 + 1$. The numbers 4, 3, 7, and 1 are called the *digits* of this number. The possible digits of a number are then the ten integers $0, 1, 2, \cdots, 9$. Actually, any positive integer greater than 1 can be used as a base in the way in which we ordinarily use 10. By this statement we mean the following.

4.7 Theorem. *Let b be a positive integer greater than 1. Then every positive integer a can be expressed uniquely in the form*

4.8 $$a = r_m b^m + r_{m-1} b^{m-1} + \cdots + r_1 b + r_0,$$

where m is a nonnegative integer and the r's are integers such that

$$0 < r_m < b \quad \text{and} \quad 0 \leq r_i < b \quad \text{for} \quad i = 0, 1, \cdots, m - 1.$$

PROOF. If 4.8 holds, it will be convenient to say that the right side of 4.8 is "a representation of a, using the base b." We shall first sketch a proof that every positive integer has such a representation, and then prove the uniqueness. In the proof we shall use the form of the Induction Principle given in Exercise 9 of Section 3.3.

If $a < b$, then 4.8 holds with $m = 0$ and $r_0 = a$; hence every positive integer less than b has a representation, using the base b. In particular, the integer 1 has such a representation. To complete the proof by induction, let us assume that every positive integer less than a has a representation, and show that a must then have a representation. By the remarks above, the case in which $a < b$ has already been disposed of, so we may assume that $a \geq b$. By the Division Algorithm, we have that

4.9 $$a = qb + r, \qquad 0 \leq r < b,$$

with $q > 0$ since $a \geq b$, and clearly $q < a$ since $b > 1$. Therefore by our assumption, q has a representation, using the base b. That is, we may write

$$q = s_k b^k + \cdots + s_1 b + s_0,$$

where k is a nonnegative integer, $0 < s_k < b$, and $0 \leq s_i < b$ for $i = 0, 1, \cdots, k - 1$. If we substitute this expression for q in Equation 4.9, we obtain

$$a = s_k b^{k+1} + \cdots + s_1 b^2 + s_0 b + r,$$

and it may be verified that this is a representation of the required form (with the m of 4.8 being $k + 1$). The Induction Principle thus shows that every positive integer has a representation, using the base b.

Let us now establish the *uniqueness* of the representation of a, using the base b. If $a < b$, a representation 4.8 must reduce merely to $a = r_0$ (that is, $m = 0$), and it is clear that there is no other representation. Accordingly, let us suppose that $a \geq b$ and for the purposes of our induction proof let us now assume that for every positive integer less than a there is a unique representation, using the base b. If 4.8 is one such representation for a, we may write

$$a = (r_m b^{m-1} + \cdots + r_1)b + r_0,$$

and since $0 \leq r_0 < b$, we see that r_0 is the remainder and that $r_m b^{m-1} + \cdots + r_1$ is the quotient in the division of a by b. Moreover, the quotient is greater than zero since $a \geq b$. Suppose, now, that in addition to 4.8 we have the following representation of a, using the base b:

$$a = t_n b^n + \cdots + t_1 b + t_0,$$

with the appropriate restrictions on these various numbers. Then by the same argument as applied above to 4.8, we see that t_0 is the remainder and $t_n b^{n-1} + \cdots + t_1$ is the quotient in the division of a by b. However, in the division algorithm the quotient and the remainder are unique. Hence $r_0 = t_0$; also we have

$$r_m b^{m-1} + \cdots + r_1 = t_n b^{n-1} + \cdots + t_1,$$

and the two sides of this equation give representations, using the base b, of a positive integer less than a. But, by our assumption, it follows that these representation must be identical. Hence we conclude that $m = n$ and that $r_i = t_i$ for $i = 1, 2, \cdots, m$. Since we already know that $r_0 = t_0$, we see that our two representations of a are identical, and the proof of uniqueness is completed by an application of the Induction Principle.

Just as we omit the powers of 10 in the usual notation, we may specify a number a, using the base b, by giving in order the "digits" $r_m, r_{m-1}, \cdots, r_1, r_0$. In order to indicate the base being used, let us specify the number a, given by 4.8, by writing $(r_m r_{m-1} \cdots r_1 r_0)_b$. If no base is indicated, it will be understood that the base is 10. For example, $(3214)_5$ really means

$$3 \cdot 5^3 + 2 \cdot 5^2 + 1 \cdot 5 + 4,$$

and it is readily verified that $(3214)_5 = 434$. The proof of the uniqueness part of the above theorem suggests an easy way to obtain the representation of a given number a, using the base b. That is, r_0 is the remainder in the division of a by b, r_1 is the remainder in the division of the preceding quotient by b, and so on.

5	434	
5	86	4
5	17	1
	3	2

We exhibit the calculations for 434, using the base 5. Here the remainders in the successive divisions are written off to the right, and the divisions are carried out until the last quotient is less than 5. We conclude, therefore, that $434 = (3214)_5$, which agrees with our previous calculations.

It is possible to carry out all the usual operations of arithmetic using entirely some fixed base other than 10. As an illustrtion, let us use base 5. In order to add or multiply any two numbers we need only learn addition and multiplication tables for the integers less than 5. In designating an integer less than 5, it is not necessary to indicate whether the base is 5 or 10. However, we have $5 = (10)_5$, $3 + 4 = (12)_5$, $3 \cdot 4 = (22)_5$, and so on. The reader may verify the addition and multiplication given below, following the usual procedure of arithmetic but using base 5 throughout.

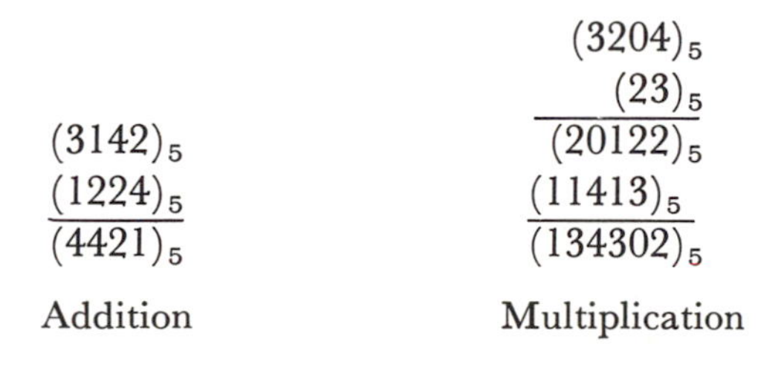

$$\begin{array}{r} (3142)_5 \\ (1224)_5 \\ \hline (4421)_5 \end{array} \qquad \begin{array}{r} (3204)_5 \\ (23)_5 \\ \hline (20122)_5 \\ (11413)_5 \\ \hline (134302)_5 \end{array}$$

Addition Multiplication

Of course, any other base can be used just as well as 5. However, the only base other than 10 that is in use to any extent is the base 2, and this system is called the *binary* system. The possible "digits" in the binary system are just 0 and 1, and expressing a number in this system involves expressing it as a sum of *different* powers of 2. For example, $(1011)_2 = 2^3 + 2 + 1$. The binary system is a most convenient one for use with many of the modern high-speed computing machines. Some of these machines are so constructed that information can be fed into the machine in the usual decimal system. The machine then expresses the given numbers in the binary system, carries out the calculations in the binary system, changes the results back into the decimal system, and automatically prints the answers.

EXERCISES

1. Write each of the following numbers using the base 5 and also using the base 2:

24, 116, 412, 3141, 2384.

2. Carry out the following additions using the indicated base:

$$\begin{array}{ccc} (1130)_5 & (2143)_5 & (101101)_2 \\ \underline{(432)_5} & \underline{(1434)_5} & \underline{(11011)_2} \end{array}$$

3. Carry out the following multiplications using the indicated base:

$$\begin{array}{ccc} (143)_5 & (4312)_5 & (10101)_2 \\ \underline{(244)_5} & \underline{(324)_5} & \underline{(1101)_2} \end{array}$$

4. Prove that every positive integer a can be expressed in the form

$$3^m + a_{m-1}3^{m-1} + \cdots + a_1 3 + a_0,$$

where m is a nonnegative integer and each of the integers $a_{m-1}, \cdots, a_1, a_0$ has the value 0, 1, or -1.

4.3 GREATEST COMMON DIVISOR

We next make the following definition.

4.10 Definition. If a and b are nonzero integers, the *greatest common divisor* (g.c.d.) of a and b is the unique *positive* integer d with the following two properties:

(i) $d|a$ and $d|b$.
(ii) If c is an integer such that $c|a$ and $c|b$, then $c|d$.

We shall presently prove the existence of the g.c.d. of two nonzero integers a and b. However, it is easy to see that if there is a positive integer which satisfies conditions (i) and (ii), it is unique. For if d and d_1 satisfy both these conditions, it follows that $d|d_1$ and $d_1|d$. Then, since they are positive, 4.2.(viii) shows that $d = d_1$.

In proving the existence of the g.c.d. we shall make use of the following concept.

4.11 Definition. If $a, b \in \mathbf{Z}$ we say that an integer of the form

$$ax + by, \qquad x,y \in \mathbf{Z},$$

is a *linear combination* of a and b.

The existence of the g.c.d., and also one of its important properties, will be established in the following theorem.

4.12 Theorem. *If a and b are nonzero integers, the least positive integer which is expressible as a linear combination of a and b is the g.c.d. of a and b. That is, if d is the g.c.d. of a and b, there exist integers x_1 and y_1 such that*

$$d = ax_1 + by_1,$$

and d is the smallest positive integer which is expressible in this form.

PROOF. For convenience of reference, let us define the set S as follows:

$$S = \{ax + by \mid x, y \in \mathbf{Z}, ax + by > 0\}.$$

Hence S is just the set of all positive integers that are expressible as linear combinations of a and b. Since, for example, $a^2 + b^2$ is an element of S, it is clear that S is not empty. Accordingly, there must exist a smallest positive integer d in the set S. Since $d \in S$, there exist integers x_1 and y_1 such that

$$d = ax_1 + by_1. \tag{4.13}$$

We next show that d is a divisor of a. By the Division Algorithm, there exist integers q and r such that

$$a = qd + r, \qquad 0 \leq r < d.$$

Using Equation 4.13, we then find that

$$\begin{aligned} r = a - qd &= a - q(ax_1 + by_1) \\ &= a(1 - qx_1) + b(-qy_1). \end{aligned}$$

It is now clear that if $r > 0$, then $r \in S$. But since $r < d$ and d is the least element of S, we conclude that $r = 0$. Hence $a = qd$, and d is a divisor of a. In a similar way it can be shown that d is a divisor of b. Hence d satisfies condition 4.10(i).

It is quite easy to see that d also satisfies condition 4.10(ii). For, suppose that c is a common divisor of a and of b, and hence that $a = a_1c$, $b = b_1c$ for certain integers a_1 and b_1. Then using Equation 4.13, we obtain

$$d = a_1cx_1 + b_1cy_1 = (a_1x_1 + b_1y_1)c,$$

and c is a divisor of d. This completes the proof of the theorem.

Although the preceding theorem establishes the existence of the g.c.d., its proof does not suggest a method for actually computing the g.c.d. of two given integers. We next present a procedure, known as the *Euclidean Algorithm*, which will be useful in this connection.

If d is the g.c.d. of the integers a and b, then, by 4.2(ii) and (iii), it is also the g.c.d. of $-a$ and b, of a and $-b$, and of $-a$ and $-b$. Accordingly, without loss of generality, we now assume that a and b are positive integers. By the Division Algorithm, we may write

$$a = qb + r, \qquad 0 \leq r < b.$$

If $r = 0$, then $b|a$ and the g.c.d. of a and b is b, so henceforth we assume that $r \neq 0$. We now divide b by r, getting

$$b = q_1 r + r_1, \qquad 0 \leq r_1 < r.$$

If $r_1 \neq 0$, we divide r by r_1, and obtain

$$r = q_2 r_1 + r_2, \qquad 0 \leq r_2 < r_1,$$

and repeat this process. Since $r > r_1 > r_2 > \cdots$, and all these remainders are nonnegative integers, we must eventually get a zero remainder. If r_{k+1} is the first zero remainder, we then have the following system of equations:

4.14

$$\begin{aligned}
a &= qb + r, & 0 &< r < b, \\
b &= q_1 r + r_1, & 0 &< r_1 < r, \\
r &= q_2 r_1 + r_2, & 0 &< r_2 < r_1, \\
r_1 &= q_3 r_2 + r_3, & 0 &< r_3 < r_2, \\
&\cdot\ \cdot\ \cdot\ \cdot\ \cdot\ \cdot\ \cdot & &\cdot\ \cdot\ \cdot\ \cdot\ \cdot \\
r_{k-2} &= q_k r_{k-1} + r_k, & 0 &< r_k < r_{k-1}, \\
r_{k-1} &= q_{k+1} r_k.
\end{aligned}$$

We now assert that r_k (the last nonzero remainder) is the g.c.d. of a and b. To establish this fact, we need to verify the two properties (i) and (ii) of Definition 4.10. First, let us show that r_k is a common divisor of a and b. We do so by starting with the last of Equations 4.14 and working back to the first as follows. It is clear from the last equation that r_k is a divisor of r_{k-1}. Since now r_k is a common divisor of r_k and r_{k-1}, the next-to-last equation shows that it is also a divisor of r_{k-2}. Proceeding in this way, when we get to the second equation we will know that r_k is a common divisor of r_1 and r, and hence is also a divisor of b. The first equation then shows that r_k is also a divisor of a. Hence, r_k is a common divisor of a and b, and 4.10(i) is established. To establish 4.10(ii), let c be any common divisor of a and b. We now use Equations 4.14 in the other order. The first equation shows that c is a divisor of r, the next that it is a divisor of r_1, and so on. Eventually, we find that it is a divisor of r_k, and the proof is complete.

Let us now give a numerical example. Suppose that we desire to compute the g.c.d. of the integers 26 and 382. By ordinary division we find that Equations 4.14 take the following form:

$$\begin{aligned} 382 &= 14\cdot 26 + 18, \\ 26 &= 1\cdot 18 + 8, \\ 18 &= 2\cdot 8 + 2, \\ 8 &= 4\cdot 2. \end{aligned}$$

In this case, 2 is the g.c.d. since it is the last nonzero remainder.

Not only is the Euclidean Algorithm useful in computing the g.c.d. of two integers, but it is also useful in expressing the g.c.d. of two integers as a linear combination of these integers. Actually *each* of the remainders in Equations 4.14 can be expressed, in turn, as a linear combination of a and b. From the first of Equations 4.14, we see that

4.15 $$r = a - qb,$$

and hence r is a linear combination of a and b. Substituting this expression for r in the second equation, and solving for r_1, we get

4.16 $$r_1 = b - q_1(a - qb) = (1 + q_1q)b - q_1a,$$

and hence r_1 is a linear combination of a and b. Now substituting from 4.15 and 4.16 into the third of Equations 4.14, we obtain

$$\begin{aligned} r_2 = r - q_2r_1 &= (a - qb) - q_2[(1 + q_1q)b - q_1a] \\ &= (1 + q_2q_1)a - (q + q_2 + q_2q_1q)b, \end{aligned}$$

so that r_2 is a linear combination of a and b. By continuing in this way, we see that each remainder and, in particular, the g.c.d. r_k is expressible as a linear combination of a and b. (See Exercise 8 below.) We are not here interested in general formulas which express these remainders as linear combinations of a and b since, in a numerical case, it is easy to compute, in turn, each of these linear combinations.

As an example, let us carry out the calculations for the case in which $a = 382$ and $b = 26$. The Euclidean Algorithm has been applied above to these two integers to find that their g.c.d. is 2. Let us now use the equations previously exhibited to express each of the remainders as a linear combination of 382 and 26. For simplicity, we shall write a in place of 382 and b in place of 26. The calculations are as follows:

$$\begin{aligned} 18 &= a - 14b, \\ 8 &= b - 18 = b - (a - 14b) = 15b - a, \\ 2 &= 18 - 2\cdot 8 = a - 14b - 2(15b - a) \\ &= 3a - 44b. \end{aligned}$$

Hence,

$$2 = 3(382) - 44(26),$$

and we have expressed the g.c.d. of 382 and 26 as a linear combination of these two integers.

We shall sometimes find it convenient to let (a, b) designate the g.c.d. of a and b. Thus, for example, we have that $(382, 26) = 2$. There can be no possible confusion with other uses of the number pair notation since the context will make it clear that we are considering the g.c.d. of two integers and not, for example, the coordinates of a point in the plane.

We shall frequently need to refer to a pair of integers with 1 as their g.c.d. Accordingly, it is convenient to make the following definition.

4.17 Definition. The integers a and b are said to be *relatively prime* if and only if their g.c.d. is 1, that is, if and only if $(a, b) = 1$.

EXERCISES

Unless otherwise specified, the letters represent arbitrary nonzero integers.

1. Find the g.c.d. of each of the following pairs of integers and express it as a linear combination of the two integers: (i) 52 and 38, (ii) 81 and 110, (iii) 320 and 112, (iv) 7469 and 2387, (v) 10,672 and -4147.

2. Show that a and b are relatively prime if and only if 1 is expressible as a linear combination of a and b.

3. If $d = (a, b)$ and $a = a_1 d$, $b = b_1 d$, show that $(a_1, b_1) = 1$.

4. If m is a positive integer, show that $(ma, mb) = m(a, b)$.

5. Show each of the following:
- (i) If p is a positive prime and a is a nonzero integer, then either $(a, p) = 1$ or $(a, p) = p$.
- (ii) If p and q are distinct positive primes, then 1 is expressible as a linear combination of p and q.

6. If $x = yz + t$, prove that $(x, z) = (z, t)$.

7. Prove that $(a, bc) = 1$ if and only if $(a, b) = 1$ and $(a, c) = 1$.

8. Write out a formal proof that every remainder in Equations 4.14 is expressible as a linear combination of a and b. [Hint: Assume that this is false, and obtain a contradiction.]

9. If a, b, and n are given, prove that n is expressible as a linear combination of a and b if and only if $(a, b)|n$.

10. (i) Define the g.c.d. of *three* nonzero integers.

(ii) Establish the existence of the g.c.d. of three integers by proving a result analogous to Theorem 4.12.

(iii) If d is the g.c.d. of a, b, and c, show that $d = ((a, b), c) = ((a, c), b) = (a, (b, c))$.

(iv) Let d be the g.c.d. of a, b, and c. If $a = a_1d$, $b = b_1d$, and $c = c_1d$, show that 1 is the g.c.d. of the three integers a_1, b_1, and c_1.

4.4 THE FUNDAMENTAL THEOREM

The principal theorem to be proved in this section has to do with the factorization of an integer into a product of primes. We begin with the following important preliminary result.

4.18 Lemma. *If a and b are nonzero integers and p is a prime such that $p|ab$, then $p|a$ or $p|b$.*

PROOF. To prove this lemma, let us suppose that p does not divide a, and show that $p|b$. Since p is a prime which is not a divisor of a, the definition of a prime implies that $(a, p) = 1$. Then Theorem 4.12 shows that there exist integers x and y such that $1 = ax + py$. Multiplying by b, we obtain

$$b = abx + bpy.$$

Since we are given that $p|ab$, clearly p divides the right member of this equation, and therefore divides b.

It is almost obvious that the preceding lemma can be generalized to apply to a product of more than two integers. For future reference we now state this more general result.

4.19 Lemma. *Let p be a prime and m an arbitrary positive integer. If $a_1, a_2, \cdots, a_m$ are nonzero integers such that $p|(a_1a_2 \cdots a_m)$, then $p|a_i$ for at least one i, $1 \leq i \leq m$.*

This lemma is easily established by induction, and the proof will be left as an exercise. The case in which $m = 2$ is covered by the preceding lemma.

It is easy to verify, for example, that $60 = 2 \cdot 2 \cdot 3 \cdot 5$, and hence that 60 can be expressed as a product of positive primes. We could also write $60 = 2 \cdot 5 \cdot 3 \cdot 2$, but we shall not consider these two factorizations as essentially different since they differ only in the order in which the prime factors are written down. With this understanding, it is true that 60 has only one factorization into a product of positive primes. This is a special case of the following important theorem.

4.20 Fundamental Theorem of Arithmetic. *Every positive integer $a > 1$ can be expressed as a product of positive primes in one and only one way (except for the order of the factors).*

In the statement of the theorem it is to be understood that, as a special case, a "product" of primes may consist of a single prime. This agreement is to take care of the case in which a is itself a prime.

PROOF. First, we shall show that every positive integer $a > 1$ can be expressed, in at least one way, as a product of positive primes. Let K be the set of all integers greater than 1 that can *not* be so expressed. If K is not the empty set, there is a least integer c in K, and clearly c is not a prime. Hence $c = c_1 c_2$, where $1 < c_1 < c$ and $1 < c_2 < c$. Since c is the least element of K, we have $c_1 \notin K$ and $c_2 \notin K$. This implies that both c_1 and c_2 can be expressed as products of primes, and since $c = c_1 c_2$ it is clear that c can also be so expressed. However, this contradicts the fact that $c \in K$, and therefore K must be the empty set. In other words, every integer $a > 1$ can be expressed as a product of positive primes.

There remains to prove the *uniqueness* of the representation of an integer as the product of primes. A good many different proofs of this fact are known, but we shall here assume that there exists an integer a which can be expressed as a product of primes in two different ways, and obtain a contradiction. Suppose, then, that $a = p_1 p_2 \cdots p_r$, and that $a = q_1 q_2 \cdots q_s$, where the p's and q's are positive primes and that the p's are not identical with the q's. From the equation

$$p_1 p_2 \cdots p_r = q_1 q_2 \cdots q_s,$$

we cancel all primes that occur on both sides and, by a suitable choice of notation, obtain an equation

4.21 $$p_1 p_2 \cdots p_i = q_1 q_2 \cdots q_j.$$

Now we must have $i \geq 1$ and $j \geq 1$ since 1 does not have a prime factor. The preceding equation then shows that $p_1 | (q_1 q_2 \cdots q_j)$ and Lemma 4.19 implies that p_1 divides some one of $q_1, q_2, \cdots, q_j$. But since the p's and q's are positive primes, we conclude that p_1 must equal some one of these q's. However, this contradicts the fact that, by the way Equation 4.21 was obtained, the same prime cannot occur on both sides of this equation. This contradiction completes the proof of the theorem.

Of course, the primes occurring in a factorization of an integer into prime factors need not all be distinct. By combining the equal primes, we see that every integer $a > 1$ can be expressed uniquely in the form

4.22 $$a = p_1^{n_1} p_2^{n_2} \cdots p_k^{n_k},$$

where the p's are distinct positive primes and each of $n_1, n_2, \cdots, n_k$ is a positive integer. The right side of 4.22 may conveniently be called the *standard form* of the integer a. As an example, $2^2 \cdot 3 \cdot 5$ is the standard form of the integer 60.

Throughout this section we have considered positive integers only. However, this is no essential restriction as we can see as follows. If $a < -1$, then $-a > 1$ and the Fundamental Theorem shows that $-a$ can be expressed uniquely as a product of positive primes. It follows that a itself is then expressible uniquely as -1 times a product of positive primes. For example, $-60 = (-1) \cdot 2^2 \cdot 3 \cdot 5$.

4.5 SOME APPLICATIONS OF THE FUNDAMENTAL THEOREM

If a and c are positive integers and c is a divisor of a, then $a = cd$ for some positive integer d. If c and d are expressed as products of prime factors, then clearly a is a product of all prime factors of c times all prime factors of d. Moreover, the Fundamental Theorem then states that this gives the unique factorization of a as a product of prime factors. It follows that the only possible prime factors of c (or of d) are the primes that are factors of a. If then a is expressed in the standard form 4.22, any divisor c of a is necessarily of the form

$$c = p_1^{m_1} p_2^{m_2} \cdots p_k^{m_k},$$

where $0 \leq m_i \leq n_i$ $(i = 1, 2, \cdots, k)$. Conversely, any integer c of this form is clearly a divisor of a.

It is now easy to obtain the g.c.d. (a, b) of two integers a and b if both a and b are expressed in standard form. Clearly, (a, b) is the product of those primes which are factors of both a and b, each such prime occurring to the smaller of the two powers to which it occurs in a and in b. For example, $60 = 2^2 \cdot 3 \cdot 5$ and $252 = 2^2 \cdot 3^2 \cdot 7$. It follows that $(60, 252) = 2^2 \cdot 3$.

We have previously had a method for computing the g.c.d. of two integers by the use of Euclid's algorithm—a method which does not involve finding any prime factors of the given integers. From a computational point of view, the previous method may involve much less work than the present one since it may be exceedingly difficult to find the prime factors of fairly large numbers, and therefore difficult to express them in standard form.

We shall not have much occasion to use the concept we now define, but we include it here for the sake of completeness.

4.23 Definition. The *least common multiple* (l.c.m.) of two nonzero integers a and b is the positive integer m with the following two properties:

(i) $a|m$ and $b|m$.
(ii) If c is an integer such that $a|c$ and $b|c$, then $m|c$.

It is easy to verify the *uniqueness* of the l.c.m., and its existence may be established in various ways. (See Exercises 3 and 5 below).

We conclude this section by giving a formal proof of the following well-known result. We shall show that there do not exist nonzero integers a and b such that

4.24 $$a^2 = 2b^2.$$

Let us suppose, on the contrary, that there do exist such integers, which we may obviously assume to be positive. If $d = (a, b)$, by Exercise 3 of the preceding set, we have $a = da_1$, $b = db_1$, where $(a_1, b_1) = 1$. Substituting in 4.24, and dividing by d^2, we find that

$$a_1^2 = 2b_1^2.$$

This equation implies that $2|a_1^2$ and Lemma 4.18 (or the Fundamental Theorem) shows that $2|a_1$. But then $4|a_1^2$ and therefore $4|2b_1^2$. This implies that $2|b_1^2$ and we must have $2|b_1$. Thus 2 is a common divisor of a_1 and b_1. We have therefore obtained a contradiction of the fact that $(a_1\ b_1) = 1$, and hence there can be no nonzero integers satisfying 4.24. The proof is therefore complete.

Another, perhaps more familiar, way of stating the result just proved is to say that $\sqrt{2}$ is not a rational number; that is, it is not expressible in the form a/b, where a and b are integers.

EXERCISES

1. Express 120 and 4851 in standard form, and find their g.c.d. and their l.c.m.

2. Do the same for 970 and 3201.

3. Explain how one can find the l.c.m. of any two integers if their standard forms are known.

4. (i) Using a method similar to that used in the proof of Lemma 4.18, show that if a is a divisor of bc and $(a, b) = 1$, then a is a divisor of c.
(ii) Prove the same result by use of the Fundamental Theorem.

5. If $a = a_1d$ and $b = b_1d$, where $d = (a, b)$, show that the l.c.m. of a and b is a_1b_1d.

6. Show that a positive integer $a > 1$ is a perfect square (that is, is the square of an integer) if and only if in the standard form of a all the exponents are even integers.

7. Show that if b and c are positive integers such that bc is a perfect square and $(b, c) = 1$, then both b and c are perfect squares.

8. Prove that there do not exist nonzero integers a and b such that $a^2 = 3b^2$.

9. If n is a positive integer which is not a perfect square, prove that there do not exist nonzero integers a and b such that $a^2 = nb^2$.

10. For each positive integer n, show that there are more than n positive primes. [Hint: Use the result of Exercise 5, Section 4.1.]

11. Prove Lemma 4.19.

★ 4.6 PYTHAGOREAN TRIPLES

If x, y, and z are *positive* integers such that

4.25
$$x^2 + y^2 = z^2,$$

we shall call the ordered triple (x, y, z) a *Pythagorean triple.* Clearly, (x, y, z) is a Pythagorean triple if and only if there exist right triangles whose sides have respective lengths x, y, and z units. Well-known examples of Pythagorean triples are (3, 4, 5), (6, 8, 10), and (5, 12, 13). In this section we shall determine all Pythagorean triples.

First we observe that we can limit our problem somewhat. If (x, y, z) is a Pythagorean triple, then so is (kx, ky, kz) for every positive integer k. Conversely, let (x, y, z) be a Pythagorean triple and suppose that d is a common divisor of x, y, and z. If we write $x = x_1d, y = y_1d$, and $z = z_1d$, we can cancel d^2 from each term of the equation

$$(x_1d)^2 + (y_1d)^2 = (z_1d)^2,$$

and find that (x_1, y_1, z_1) is also a Pythagorean triple. If it happens that d is the g.c.d. of the *three* integers x, y, and z (see Exercise 10, Section 4.3), then x_1, y_1, and z_1 have 1 as their g.c.d. Let us say that a Pythagorean triple (a, b, c) is a *primitive* Pythagorean triple if a, b, and c have 1 as their g.c.d. Then the observations that we have just made assure us that *every* Pythagorean triple is of the form (ra, rb, rc), where (a, b, c) is a primitive Pythagorean triple and r is a positive integer. Our general problem is therefore reduced to the problem of finding all primitive Pythagorean triples.

Now let (x, y, z) be a primitive Pythagorean triple. It is easy to see that *each pair* of the numbers x, y, and z must be relatively prime. If, on the contrary, two of these numbers were not relatively prime, they would have a common prime factor p. Then Equation 4.25 would show that p is also a factor of the third, which would contradict the assumption that (x, y, z) is primitive. As a special case of what we have just proved, we see that x and y cannot both be even. We next show that, also, x and y cannot both be odd. If they were both odd, we could write $x = 2m + 1$ and $y = 2n + 1$, where m and n are properly chosen integers. But then we would have

$$\begin{aligned} z^2 &= x^2 + y^2 = (2m + 1)^2 + (2n + 1)^2 \\ &= 2(2m^2 + 2n^2 + 2m + 2n + 1). \end{aligned}$$

Since the second factor in this last expression is odd, we see that z^2 would be divisible by 2 but not by 4, and this is clearly impossible. It follows that x and y cannot both be odd. We have therefore proved that one of the integers x and y must be even and the other odd. It is trivial that (a, b, c) is a primitive Pythagorean triple if and only if (b, a, c) is also, and there will be no real loss of generality if we now limit ourselves to the study of primitive Pythagorean triples (x, y, z) in which x is even, and therefore y is odd.

We are now ready to prove the following theorem.

4.26 Theorem. *If (x, y, z) is a primitive Pythagorean triple in which x is even, then*

4.27 $$x = 2uv, \quad y = u^2 - v^2, \quad z = u^2 + v^2,$$

where u and v are positive integers satisfying the following three conditions:

(i) u and v are relatively prime.
(ii) $u > v$.
(iii) One of u, v is even and the other is odd.

Conversely, if u and v are any positive integers satisfying these three conditions and x, y, and z are determined by Formulas 4.27, then (x, y, z) is a primitive Pythagorean triple in which x is even.

PROOF. To prove the first part of the theorem, let (x, y, z) be a primitive Pythagorean triple in which x is even. We have proved that no two of x, y, and z can be even; hence y and z are both odd. This implies that $z + y$ and $z - y$ are both even; that is, that there exist positive integers r and s such that

4.28 $$z + y = 2r, \quad z - y = 2s.$$

From these, it is easy to verify that

4.29 $$z = r + s, \quad y = r - s.$$

Now r and s must be relatively prime since any common factor of r and s would be a common factor of the relatively prime integers z and y. Moreover, since

$$x^2 = z^2 - y^2 = (z + y)(z - y),$$

it follows from Equations 4.28 that

4.30 $$x^2 = 4rs.$$

Since x is even, $x = 2t$ for some integer t, and the preceding equation shows that

4.31 $$t^2 = rs.$$

Since r and s are relatively prime, Exercise 7 of the preceding set implies that both r and s are perfect squares. That is, there exist positive integers u and v such that

4.32 $$r = u^2, \quad s = v^2.$$

Then Equations 4.30 and 4.29 show that

4.33 $$x = 2uv, \quad y = u^2 - v^2, \quad z = u^2 + v^2,$$

and Formulas 4.27 are satisfied. There remains only to prove that u and v have the required properties. Since r and s are relatively prime, it follows from 4.32 that u and v are relatively prime. Next, we see that $u > v$ since $y > 0$. We already know that u and v cannot both be even inasmuch as they are relatively prime. Finally, from 4.33 it follows that they cannot both be odd since otherwise y (and z also) would be even, whereas we know that it is odd. This completes the proof of the first part of the theorem.

To prove the second part, suppose that u and v are any positive integers satisfying conditions (i), (ii), and (iii); and let x, y, and z be defined by Formulas 4.27. Clearly x, y, and z are all positive, and it is easy to verify that

$$(2uv)^2 + (u^2 - v^2)^2 = (u^2 + v^2)^2,$$

and hence that (x, y, z) is a Pythagorean triple. We shall prove that it is necessarily primitive by showing that y and z are relatively prime. By condition (iii), $u^2 - v^2$ and $u^2 + v^2$ are both odd, that is, y and z are both odd. If y and z were not relatively prime, they would have a common prime factor $p \neq 2$. But since $z + y = 2u^2$ and $z - y = 2v^2$, it would follow that p is also a common factor of u and v. However, it is given that u and v are relatively prime, and therefore y and z can have no common prime factor. Hence, (x, y, z) is a primitive Pythagorean triple. It is obvious that x is even, and the proof is therefore complete.

It follows from the theorem that there are infinitely many primitive Pythagorean triples. The triple (4, 3, 5) is obtained by setting $u = 2$, $v = 1$ in 4.27; the triple (12, 5, 13) by choosing $u = 3$, $v = 2$; the triple (8, 15, 17) by choosing $u = 4$, $v = 1$; and so on.

4.7 THE RING OF INTEGERS MODULO n

In this section we shall make use of the concepts of equivalence relation and equivalence set, which were defined in Chapter 1. First, we make the following definition.

4.34 Definition. Let n be a fixed integer greater than 1. If a and b are integers such that $a - b$ is divisible by n, we say that "a is congruent to b modulo n," and indicate this by writing $a \equiv b \pmod{n}$.

As an illustration of the use of this notation, let $n = 5$. Then we have $18 \equiv 3 \pmod 5$ since $18 - 3$ is divisible by 5. In like manner, $-2 \equiv 8 \pmod 5$, $4 \equiv 4 \pmod 5$, $1342 \equiv 2 \pmod 5$, and so on.

We leave it to the reader to verify that congruence modulo n is an equivalence relation on the set **Z** of all integers. By this we mean, of course, that the three properties of Definition 1.8 of an equivalence relation are satisfied. We may emphasize that throughout this section n will always be a positive integer greater than 1.

Now that we have an equivalence relation on the set **Z**, we can consider equivalence sets as introduced in Definition 1.9. We may point out that, relative to the equivalence relation of congruence modulo n, an equivalence set $[k]$ is defined as follows:

$$[k] = \{x \mid x \in \mathbf{Z}, x \equiv k \pmod n\}.$$

For convenience, we shall refer to $[k]$ as an "equivalence set modulo n." If $[k]$ and $[l]$ are equivalence sets modulo n, then 1.11(i) shows that $[k] = [l]$ if and only if $k \equiv l \pmod n$.

Next, let us observe that if $a \in \mathbf{Z}$ and r is the remainder in the division of a by n, then necessarily $a \equiv r \pmod n$. This follows from the observation that if $a = qn + r$, then $a - r = qn$ and hence $a \equiv r \pmod n$. Since we have $0 \leq r < n$, it follows that every integer is congruent modulo n to some one of the n integers $0, 1, 2, \cdots, n - 1$. Moreover, since each of these integers is less than n, no two of them can be congruent to each other modulo n. Since $[k] = [l]$ if and only if $k \equiv l \pmod n$, we have shown that there are precisely n different equivalence sets modulo n, namely, the sets $[0], [1], [2], \cdots, [n - 1]$.

As an example, let us take $n = 5$. It is easy to verify that the five equivalence sets modulo 5 are the following:

4.35
$$\begin{aligned}
[0] &= \{\cdots, -15, -10, -5, 0, 5, 10, 15, \cdots\}, \\
[1] &= \{\cdots, -14, -9, -4, 1, 6, 11, 16, \cdots\}, \\
[2] &= \{\cdots, -13, -8, -3, 2, 7, 12, 17, \cdots\}, \\
[3] &= \{\cdots, -12, -7, -2, 3, 8, 13, 18, \cdots\}, \\
[4] &= \{\cdots, -11, -6, -1, 4, 9, 14, 19, \cdots\}.
\end{aligned}$$

It follows easily from 1.11(i) that if $a \in [2]$, then necessarily $[a] = [2]$. Hence, $[2] = [-3] = [17]$, and so on. In view of the observation made above, we see that $[2]$ consists of all those integers a such that the remainder in the division of a by 5 is 2. Otherwise expressed, the equivalence set $[2]$ consists of the set of all integers of the form $5q + 2$, $q \in \mathbf{Z}$. Similar remarks hold for the other equivalence sets modulo 5, as well as for equivalence sets modulo n.

Still considering the special case of congruence modulo 5, let us give a description of the procedure, which we shall justify in detail below, by which we propose to construct a new ring. Let $U = \{[0], [1], [2], [3], [4]\}$, and hence an element of the set U is just one of the equivalence sets 4.35. We propose to make U into a ring by suitable definitions of addition and multiplication of its elements. What shall we mean, for example, by $[2] + [4]$? By examining 4.35 it appears that the sum of an element of [2] and an element of [4] always gives an element of [6]. Hence, it is natural to define $[2] + [4] = [6]$. Of course, $[6] = [1]$, so we could equally well say that $[2] + [4] = [1]$. Similarly, the product of an element of [2] and an element of [4] is always an element of [8], and we therefore define $[2] \cdot [4] = [8]$. Again, since $[8] = [3]$, this is the same as saying that $[2] \cdot [4] = [3]$. In a similar way we could define the sum or the product of any two elements of U, always obtaining an element of U. The importance of all this is that with respect to these operations of addition and multiplication on U, it can be shown that U is a ring. This ring we shall call the "ring of integers modulo 5," and denote it by $\mathbf{Z}_5$. The ring $\mathbf{Z}_5$ therefore has five elements, each element being one of the equivalence sets 4.35.

We now proceed to justify these statements, and to generalize them to the case of congruence modulo n. The following properties of congruence modulo n are fundamental for our purpose.

4.36 Theorem. *If $a \equiv b \pmod{n}$ and $c \equiv d \pmod{n}$, then*

(i) $$a + c \equiv b + d \pmod{n},$$

(ii) $$ac \equiv bd \pmod{n}.$$

PROOF. To prove these properties, we observe that $a \equiv b \pmod{n}$ means that there is an integer k such that $a = b + kn$. Similarly, we have $c = d + ln$ for some integer l. It follows that

$$a + c = b + d + (k + l)n,$$

that is, that $a + c \equiv b + d \pmod{n}$, and (i) is established. The second part follows easily by observing that

$$\begin{aligned} ac &= (b + kn)(d + ln) \\ &= bd + (bl + kd + kln)n, \end{aligned}$$

and hence $ac \equiv bd \pmod{n}$, as required.

For the moment, let us denote by T the set of all equivalence sets modulo n. We proceed to define operations of addition and multiplication on T as follows. If $[r]$, $[s] \in T$, we define

4.37 $$[r] + [s] = [r + s]$$

and

4.38 $$[r] \cdot [s] = [rs].$$

Now in order to show that these, in fact, do define addition and multiplication on the set T, we must show that the sum and product of the equivalence sets $[r]$ and $[s]$ do not depend upon the particular *notation* used to designate these sets, but only upon the sets themselves. This is sometimes expressed by saying that we must show that addition and multiplication are *well-defined* by 4.37 and 4.38. To clarify this statement, for the moment let us again consider the case in which $n = 5$. Then, by 4.37, we have that $[6] + [2] = [8]$ and that $[1] + [-3] = [-2]$. So far, the fact that $n = 5$ has played no role. But now, using this value of n, we may observe that $[6] = [1]$ and that $[2] = [-3]$, so we would certainly want to have $[6] + [2] = [1] + [-3]$. Since, in fact, $[8] = [-2]$, we see that this is indeed the case. The fact that a similar result always holds is what we mean by saying that addition of equivalence sets is well-defined. Now let n be arbitrary and let us show that the preceding theorem gives us exactly the information which we need at this point. Suppose that x, $r \in \mathbf{Z}$ such that $[x] = [r]$, and that y, $s \in \mathbf{Z}$ such that $[y] = [s]$. These imply that $x \equiv r \pmod{n}$ and that $y \equiv s \pmod{n}$. The theorem then asserts that $x + y \equiv r + s \pmod{n}$ and that $xy \equiv rs \pmod{n}$, that is, that $[x + y] = [r + s]$ and that $[xy] = [rs]$. Hence, addition and multiplication of elements of T are indeed well-defined by 4.37 and 4.38.

It is now quite easy to establish the following result.

4.39 Theorem. *With respect to the definitions 4.37 and 4.38 of addition and multiplication, the set of all equivalence sets of* $\mathbf{Z}$ *modulo* n *is a commutative ring with unity. This ring is called "the ring of integers modulo* n*," and denoted by* $\mathbf{Z}_n$.

PROOF. Let us prove, for example, the associative law of addition for elements of $\mathbf{Z}_n$. If $[r]$, $[s]$, and $[t]$ are elements of $\mathbf{Z}_n$, we wish therefore to prove that

4.40 $$([r] + [s]) + [t] = [r] + ([s] + [t]).$$

Now by 4.37, $[r] + [s] = [r + s]$, and again applying 4.37, we see that the left side of 4.40 is the element $[(r + s) + t]$ of $\mathbf{Z}_n$. A similar calculation shows that the right side of 4.40 is equal to $[r + (s + t)]$. However, by the associative law of addition *for the integers*, we know that $(r + s) + t = r + (s + t)$, and it follows that 4.40 must hold.

In a similar way, each of the other properties of $\mathbf{Z}_n$ which need to be verified in order to establish the theorem follows from the corresponding property of the ring $\mathbf{Z}$. The proofs of these will be left as exercises. In particular, it is almost obvious that [0] is the zero of the ring $\mathbf{Z}_n$ and that [1] is its unity.

As an illustration, let us again consider the special case in which $n = 5$. As pointed out above, the five elements of this ring are [0], [1], [2], [3], and [4]. The reader may verify the following addition and multiplication tables for this ring. For convenience, we have omitted the brackets and written "k" in place of "$[k]$." This is often done when the context makes the meaning clear.

$(+)$	0	1	2	3	4
0	0	1	2	3	4
1	1	2	3	4	0
2	2	3	4	0	1
3	3	4	0	1	2
4	4	0	1	2	3

$(\cdot)$	0	1	2	3	4
0	0	0	0	0	0
1	0	1	2	3	4
2	0	2	4	1	3
3	0	3	1	4	2
4	0	4	3	2	1

The Ring $\mathbf{Z}_5$

By examining the multiplication table for this ring we see that this ring has no nonzero divisors of zero, and hence that the ring is an integral domain.. The following theorem tells us for just what integers n the ring $\mathbf{Z}_n$ is an integral domain.

4.41 Theorem. *The ring $\mathbf{Z}_n$ is an integral domain if and only if n is a prime.*

PROOF. First, suppose that n is a prime p, and that $[r]$ and $[s]$ are elements of $\mathbf{Z}_p$ such that $[r] \cdot [s] = [0]$. Then $rs \equiv 0 \pmod{p}$, which implies that rs is divisible by p. Now since p is a prime, it follows that r is divisible by p or s is divisible by p, that is, that $r \equiv 0 \pmod{p}$ or $s \equiv 0 \pmod{p}$. Hence, $[r] = [0]$ or $[s] = [0]$, and $\mathbf{Z}_p$ is an integral domain.

Now suppose that n is not prime. It follows that there exist integers n_1 and n_2 such that $n = n_1n_2$, $1 < n_1 < n$, $1 < n_2 < n$. Hence, in $\mathbf{Z}_n$, we have $[n_1]\cdot[n_2] = [0]$, with $[n_1] \neq [0]$ and $[n_2] \neq [0]$. This shows that the ring $\mathbf{Z}_n$ is not an integral domain since it does have nonzero divisors of zero.

Although we have not as yet mentioned this important fact, let us now point out that the definitions of addition and multiplication of equivalence sets modulo n show that *the ring* $\mathbf{Z}_n$ *is a homomorphic image of the ring* $\mathbf{Z}$. More precisely, the mapping $\theta: \mathbf{Z} \to \mathbf{Z}_n$ defined by $r\theta = [r]$, $r \in \mathbf{Z}$, is a homomorphic mapping of $\mathbf{Z}$ onto $\mathbf{Z}_n$. Indeed, it is this fact which assures us that in the proof of Theorem 4.39 each of the ring properties in $\mathbf{Z}_n$ is a consequence of the corresponding property in the ring $\mathbf{Z}$. (See Exercise 12 below.)

EXERCISES

1. Prove that congruence modulo n is an equivalence relation on the set $\mathbf{Z}$.
2. If $a \equiv b \pmod{n}$, prove by induction that $a^m \equiv b^m \pmod{n}$ for every positive integer m.
3. Complete the proof of Theorem 4.39.
4. Let S be the set of all positive integers n such that $n > 1$. If $a, b \in S$, let us define $a \sim b$ to mean that a and b have the same number of positive prime factors (distinct or identical). Show that "$\sim$" is an equivalence relation defined on S. If $[a]$ is the equivalence set (relative to this equivalence relation) which contains the integer a, and we set $[a] + [b] = [a + b]$, verify that we do *not* have a well-defined addition of equivalence sets.
5. Make addition and multiplication tables for each of the following rings: $\mathbf{Z}_2$, $\mathbf{Z}_3$, $\mathbf{Z}_4$, $\mathbf{Z}_6$, and $\mathbf{Z}_7$. By examining the multiplication tables, determine which of these rings are integral domains and compare with Theorem 4.41.
6. Verify that the ring $\mathbf{Z}_2$ is isomorphic to the ring of Example 5 of Section 2.3.
7. Determine whether the ring $\mathbf{Z}_4$ is isomorphic to either of the rings of Examples 6 and 7 of Section 2.3.
8. Verify that the elements $[0]$, $[3]$, $[6]$, and $[9]$ of the ring $\mathbf{Z}_{12}$ are the elements of a subring of $\mathbf{Z}_{12}$. Find an isomorphic mapping of $\mathbf{Z}_4$ onto this subring.
9. Is the integral domain $\mathbf{Z}_p$ (p a prime) an ordered integral domain? Why?

10. Characterize those integers a such that the element $[a]$ of $\mathbf{Z}_n$ is a divisor of zero.

11. Prove that the nonzero element $[a]$ of the ring $\mathbf{Z}_n$ has a multiplicative inverse in $\mathbf{Z}_n$ if and only if a and n are relatively prime.

12. Let R be a ring and S a set on which operations of addition and multiplication are defined. If there exists a mapping θ of R onto S which preserves the operations of addition and multiplication, prove that S is a ring with respect to the operations of addition and multiplication defined on S. (This fact is often expressed by saying that "a homomorphic image of a ring is a ring.")

13. If $[a]$ is the equivalence set relative to congruence modulo n which contains a, prove that if $x, y \in [a]$, then $(x, n) = (y, n)$.

14. Prove that if $(i, n) = 1$, the mapping $\theta\colon \mathbf{Z}_n \to \mathbf{Z}_n$ defined by $[a]\theta = i[a]$ is a one-one mapping of $\mathbf{Z}_n$ onto $\mathbf{Z}_n$. Is it an isomorphism?

15. An element a of a ring R is said to be *idempotent* if $a^2 = a$. If m and n are relatively prime positive integers greater than 1, prove that the ring $\mathbf{Z}_{mn}$ has at least two idempotents other than the zero and the unity. [Hint: If $1 = mx + ny$, consider $[mx]$ and $[ny]$.]

16. (i) If $n = 2k$, where k is a positive odd integer, verify that $k^2 \equiv k \pmod{n}$.
    (ii) If n is as in the preceding, prove that the ring $\mathbf{Z}_n$ contains a subring which is isomorphic to $\mathbf{Z}_2$. (A more general result is asked for in Exercise **18** below.)

17. Let m and n be positive integers greater than 1, and let us denote elements of $\mathbf{Z}_{mn}$ and of $\mathbf{Z}_n$ by subscripts mn and n, respectively. Verify that the mapping $\theta\colon \mathbf{Z}_{mn} \to \mathbf{Z}_n$ defined by $[i]_{mn}\theta = [i]_n$, $i \in \mathbf{Z}$, is well-defined (that is, that it actually defines a mapping of $\mathbf{Z}_{mn}$ into $\mathbf{Z}_n$) and that it is a homomorphic mapping of $\mathbf{Z}_{mn}$ onto $\mathbf{Z}_n$.

18. If m and n are relatively prime positive integers greater than 1, prove that the ring $\mathbf{Z}_{mn}$ has a subring which is isomorphic to the ring $\mathbf{Z}_n$. [Hint: If $am + bn = 1$, show that the mapping ϕ defined by $[iam]_{mn}\phi = [i]_n$, $i \in \mathbf{Z}$, is well-defined and is the desired isomorphism.]

NOTES AND REFERENCES

This chapter has been a brief introduction to what is called *The Theory of Numbers*. The word "numbers" used in this connection usually means "integers." A few books on the subject are items [18] through [23] of the Bibliography. Davenport [18] and Ore [22] are particularly recommended as being interesting and easy to read. Hardy and Wright [19] and Sierpinski [23] treat a very wide range of topics. Many other books can be found by examining the bibliographies in these or other books on the subject.

Chapter 5

Fields and the Rational Numbers

The ring **Z** of integers has the property, which is not true of the natural numbers (positive integers) alone, that every equation of the form $a + x = b$, where $a, b \in \mathbf{Z}$, has a solution x in **Z**. In fact, one of the principal reasons for the introduction of the negative integers and zero is to assure us that every such equation is solvable. However, in **Z**, an equation of the form $ax = b$ is solvable if and only if a is a divisor of b. Clearly, in order that an equation of this form, where $a, b \in \mathbf{Z}$ and $a \neq 0$, always have a solution we need to have available the rational numbers as well as the integers. Later on in this chapter we shall show how to extend the ring **Z** of integers to the larger system of rational numbers. In this process we shall not use any previous knowledge of the rational number system, except perhaps to motivate the procedure used, but shall carry out the construction using only the properties of the integers which have already been given. Before presenting this construction we shall introduce and discuss the important concept of a *field.*

In a final optional section we shall briefly indicate how the ring of integers can be constructed from the system of natural numbers. This material is presented here because the method closely parallels that by which the rational numbers are constructed from the integers.

5.1 FIELDS

Let us make the following definition.

5.1 Definition. A commutative ring F with more than one element and having a unity is said to be a *field* if every nonzero element of F has a multiplicative inverse in F.

In view of Theorem 2.10, we know that every nonzero element of a field has a *unique* multiplicative inverse. As indicated in Section 2.5, we may denote the multiplicative inverse of a nonzero element r of a field F by r^{-1}. If 1 is the unity of F, r^{-1} is therefore the unique element of F such that

5.2 $$r \cdot r^{-1} = r^{-1} \cdot r = 1.$$

However, the commutative law of multiplication is required to hold in a field, and we shall henceforth use it without explicit mention. In particular, we may consider that r^{-1} is defined by the single equation $r \cdot r^{-1} = 1$.

We shall now prove the following result.

5.3 Theorem. *A field is necessarily an integral domain.*

PROOF. Suppose that r and s are elements of a field F such that $rs = 0$. If $r \neq 0$, r has a multiplicative inverse r^{-1} in F and it follows that

$$r^{-1}(rs) = (r^{-1}r)s = 1 \cdot s = s.$$

But also,

$$r^{-1}(rs) = r^{-1} \cdot 0 = 0.$$

Hence $s = 0$, and we have shown that $r = 0$ or $s = 0$. This proves that F has no nonzero divisors of zero and F therefore satisfies Definition 3.3 of an integral domain.

Although in the definition of a field we only required the existence of the multiplicative inverse of each nonzero element r, that is, the solvability of each equation of the form $rx = 1$, we can easily establish the following more general result.

5.4 Theorem. *If r and s are elements of a field F and $r \neq 0$, there exists a unique element y of F such that $ry = s$. Moreover, $y = r^{-1} \cdot s$.*

PROOF. It is clear that $r^{-1} \cdot s$ *is* a solution of this equation since

$$r(r^{-1} \cdot s) = (r \cdot r^{-1})s = 1 \cdot s = s.$$

To obtain the *uniquness* of the solution, suppose that $ry_1 = s$ and $ry_2 = s$. Then $ry_1 = ry_2$ and, since $r \neq 0$, the cancellation law of multiplication shows that $y_1 = y_2$.

Let us now give a few examples of fields. It is to be understood that the usual definitions of addition and multiplication are implied.

Example 1: The set of all rational numbers; that is, all numbers of the form a/b, where $a, b \in \mathbf{Z}$ with $b \neq 0$.

Example 2: The set of all real numbers of the form $x + y\sqrt{2}$, where x and y are rational numbers. What is the multiplicative inverse of each nonzero element?

Example 3: The set of all real numbers of the form $u + v\sqrt{3}$, where u and v are elements of the field of the preceding example. It is true that every nonzero element has a multiplicative inverse in this set, but we shall not here write out a proof of this fact.

The following theorem gives some other examples of fields of quite a different type.

5.5 Theorem. *If p is a prime, the ring $\mathbf{Z}_p$ of integers modulo p is a field.*

This theorem is implied by the result of Exercise 11 of the preceding set of exercises. However, we now give a detailed proof of this important theorem.

PROOF. We already know that $\mathbf{Z}_p$ is a commutative ring with unity and, using the notation of Section 4.7, the unity is $[1]$. Let $[r]$ be any nonzero element of $\mathbf{Z}_p$. In order to show that we have a field, we need to show that there exists an element $[x]$ of $\mathbf{Z}_p$ such that $[r] \cdot [x] = [1]$. The fact that $[r] \neq [0]$ implies that $r \not\equiv 0 \pmod{p}$, that is, that r is not divisible by p. Since p is a prime, it follows that r and p are relatively prime and Theorem 4.12 then assures us that there exist integers x, y such that $rx + py = 1$. This implies that $rx \equiv 1 \pmod{p}$ and hence that $[r] \cdot [x] = [1]$, as required.

We may remark that if n is not prime, we know by Theorem 4.41 that the ring $\mathbf{Z}_n$ is not even an integral domain, and certainly then is not a field.

A field $\mathbf{Z}_p$ differs from the usual fields of elementary algebra in that it has only a finite number of elements. However, in this as in any field we can always perform the so-called rational operations of addition multiplication, subtraction, and division (except by zero). We are here using the familiar word "subtraction" to mean addition of additive inverse and "division" to mean multiplication by the multiplicative inverse. We may emphasize that division by zero is not defined in any field F since $0 \cdot x \neq 1$ for every x in F, and therefore 0 cannot have a multiplicative inverse.

We have required that a field contain at least two elements, and we may now observe that there does exist a field having exactly two elements, namely, the field $\mathbf{Z}_2$.

In view of Theorems 4.41 and 5.5, we know that if the ring $\mathbf{Z}_n$ is an integral domain, it is actually a field. On the other hand, the ring $\mathbf{Z}$ of integers is an integral domain which is not a field. In this connection, the following theorem may be of some interest.

5.6 Theorem. *An integral domain S with a finite number of elements is necessarily a field.*

PROOF. Suppose that S has exactly n distinct elements $a_1, a_2, \cdots, a_n$ and, for later convenience, let us agree that a_1 is the unity of S. Now suppose that a_k is an arbitrary nonzero element of S and let us show that it has a multiplicative inverse in S. Consider the set $A = \{a_ka_1, a_ka_2, \cdots, a_ka_n\}$ consisting of the product of the n elements of S by the element a_k. Since $a_k \neq 0$, the cancellation law of multiplication shows that $a_ka_i = a_ka_j$ only if $a_i = a_j$. Hence no two of the indicated elements of A can be equal. That is, the elements of A are n distinct elements of S and therefore are *all* the elements of S. Since one of them is the unity a_1, there must exist some element a_l of S such that $a_ka_l = a_1$, and a_l is the multiplicative inverse of a_k. This argument shows that every nonzero element of S has a multiplicative inverse in S and therefore that S is a field.

Incidentally, the proof of this theorem furnishes an alternate proof of Theorem 5.5 since we already know by Theorem 4.41 that $\mathbf{Z}_p$ is an integral domain (with a finite number of elements) if p is a prime.

EXERCISES

1. Find the multiplicative inverse of each nonzero element of each of the following fields: $\mathbf{Z}_5$, $\mathbf{Z}_7$, $\mathbf{Z}_{11}$, $\mathbf{Z}_{13}$, $\mathbf{Z}_{17}$.
2. Find the multiplicative inverse of each of the following elements of the field $\mathbf{Z}_{1847}$: [12], [35], [416], [800].
3. Show that there exists a finite field which is not isomorphic to $\mathbf{Z}_p$ for any prime p. [Hint: Consider Example 7 of Section 2.3.]
4. Determine at exactly what point the proof of Theorem 5.6 makes use of the assumption that S has a finite number of elements. [Hint: Try to apply the proof to the integral domain $\mathbf{Z}$ and see where it breaks down.]
5. On the set $\mathbf{Q} \times \mathbf{Q}$, $\mathbf{Q}$ the set of rational numbers, let us define addition and multiplication as follows:

$$(a, b) + (c, d) = (a + c, b + d),$$
$$(a, b) \cdot (c, d) = (ad + bc, 2ac + bd).$$

Verify that we thus get a field F. Find a subfield of F which is isomorphic to the field $\mathbf{Q}$.

6. Prove: If R is a commutative ring with more than one element and with the property that for $a, b \in R$, $a \neq 0$, there exists $x \in R$ such that $ax = b$, then R is a field.
7. Let P be the set of all positive real numbers, and q a fixed positive real number not equal to 1. If $a, b \in P$, we define operations of addition "$\oplus$" and multiplication "$\odot$" on P as follows:

$$a \oplus b = ab \text{ (ordinary multiplication)}, \quad a \odot b = a^{\log_q b}.$$

Assuming as known all the familiar elementary properties of logarithms, prove that P is a field with respect to these definitions of addition and multiplication.

5.2 THE CHARACTERISTIC

Although we are now primarily interested in fields, the concept to be introduced in this section applies to any ring and we therefore give the definition in its general form. We recall that if a is an element of a ring and n is a positive integer, we have given in Section 2.6 a recursive definition of na. We now make the following definition.

5.7 Definition. Let R be a ring. If there exists a positive integer n such that $na = 0$ for every element a of R, the smallest such positive integer n is called the *characteristic* of R. If no such positive integer exists, R is said to have *characteristic zero.*

All the familiar number systems of elementary algebra certainly have characteristic zero. However, let us consider, for example, the ring $\mathbf{Z}_4$ of integers modulo 4. If $[r]$ is any element of this ring, then $2[r] = [r] + [r] = [2r]$ and, generally, if k is a positive integer, $k[r] = [kr]$. The smallest positive integer k such that $[kr] = [0]$ for every element $[r]$ of $\mathbf{Z}_4$ is clearly 4, so $\mathbf{Z}_4$ has characteristic 4. In general, the ring $\mathbf{Z}_n$ has characteristic n.

The definition of the characteristic of a ring makes an assertion about *every* element of the ring. However, in an important special case, the following theorem shows that the characteristic is determined by some one particular element.

5.8 Theorem. *Let R be a ring with a unity e. If there exists a positive integer n such that $ne = 0$, then the smallest such positive integer is the characteristic of R. If no such positive integer exists, then R has characteristic zero.*

PROOF. If n is the smallest positive integer such that $ne = 0$, the characteristic of R certainly cannot be a positive integer less than n. Moreover, if $a \in R$, then

$$na = (na)e = (ne)a = 0a = 0,$$

so that $na = 0$ for every element a of R; hence R has characteristic n. The last sentence of the theorem is an immediate consequence of the definition of characteristic zero.

We know that the ring $\mathbf{Z}_n$ is a field if and only if n is a prime. Hence the characteristic of every field that has been mentioned so far is either zero or a prime. In fact, we shall now prove that this is always true for every integral domain and certainly then for every field.

5.9 Theorem. *Every integral domain D has characteristic zero or a prime.*

PROOF. To prove this theorem, suppose that D has characteristic $n > 0$, and that n is *not* a prime. Then $n = n_1 n_2$, where $1 < n_1 < n$, $1 < n_2 < n$. If e is the unity of D, we have $ne = 0$ and therefore $(n_1 n_2)e = 0$. However, this implies that $(n_1 e)(n_2 e) = 0$ and, by the definition of an integral domain, it follows that $n_1 e = 0$

or $n_2e = 0$. But if $n_1e = 0$, the preceding theorem shows that D cannot have characteristic $n > n_1$; hence $n_1e \neq 0$. Similarly, $n_2e \neq 0$, and we have a contradiction of the assumption that n is not a prime. The proof is therefore complete.

We next consider the following result.

5.10 Theorem. *An integral domain with characteristic zero has a subring which is isomorphic to* $\mathbf{Z}$. *An integral domain with characteristic a prime* p *has a subring which is isomorphic to* $\mathbf{Z}_p$.

We shall present some of the steps in the proof of this theorem and leave some of the details to the reader.

PROOF. Let D be an integral domain with unity e, and let us set

$$\mathbf{Z}e = \{ne \mid n \in \mathbf{Z}\}.$$

Then $\mathbf{Z}e$ is a subring of D and it is this subring which we consider.

First, suppose that D has characteristic zero and let $\theta: \mathbf{Z} \to \mathbf{Z}e$ be the mapping defined by $n\theta = ne$, $n \in \mathbf{Z}$. Clearly, θ is an onto mapping, so let us verify that it is a one-one mapping. If $n_1e = n_2e$ with $n_1, n_2 \in \mathbf{Z}$, the notation can be so chosen that $n_1 \geq n_2$. Then $(n_1 - n_2)e = 0$ and, by Theorem 5.8, we cannot have $n_1 - n_2 > 0$ since this would violate the assumption that D has characteristic zero. Accordingly, we conclude that $n_1 = n_2$ and θ is indeed a one-one mapping. It is then easily verified that addition and multiplication are preserved under the mapping θ, and hence that θ is an isomorphism of $\mathbf{Z}$ onto $\mathbf{Z}e$.

Next, suppose that D has characteristic p and let us, as usual, denote the equivalence set modulo p which contains the integer n by $[n]$. Let $\phi: \mathbf{Z}_p \to \mathbf{Z}e$ be the mapping defined by $[n]\phi = ne$, $n \in \mathbf{Z}$. We leave it to the reader to verify that this mapping is well-defined. However, let us show that it is a one-one mapping. Suppose that $n_1, n_2 \in \mathbf{Z}$ such that $n_1e = n_2e$ or $(n_1 - n_2)e = 0$. By the Division Algorithm.

$$n_1 - n_2 = qp + r, \qquad 0 \leq r < p.$$

Thus

$$(n_1 - n_2)e = (qp + r)e = qpe + re = re.$$

This shows that $re = 0$ and, by Theorem 5.8, it cannot be true that r is a positive integer less than p. Hence we must have $r = 0$

and therefore $n_1 - n_2 = qp$. It follows that $[n_1] = [n_2]$ and ϕ is thus a one-one mapping. We leave it to the reader to verify that ϕ is the desired isomorphism of $\mathbf{Z}_p$ onto the subring $\mathbf{Z}e$ of D.

The literature on the theory of fields is quite extensive. In much of this work the concept of the characteristic of a field plays an essential role. It frequently happens, for example, that although a theorem may be true for every field, different proofs have to be given for the case in which the characteristic is zero and that in which it is a prime. Later on in this book we shall present additional examples of fields.

5.3 SOME FAMILIAR NOTATION

Let F be a field with unity 1, and t a nonzero element of F. We have introduced the symbol t^{-1} to designate the multiplicative inverse of t, and have found that if $s \in F$, the unique element x of F such that $tx = s$ is given by $x = t^{-1}s$. In accordance with familiar usage, we shall also designate this element $t^{-1}s$ by $\frac{s}{t}$ or by s/t. In particular, we have $t^{-1} = 1/t$.

Suppose, now, that v is also a nonzero element of F. Since $(tv)(v^{-1}t^{-1}) = 1$, the multiplicative inverse of tv is $v^{-1}t^{-1}$, that is,

5.11 $$(tv)^{-1} = v^{-1}t^{-1}.$$

It is now easy to see that

5.12 $$\frac{sv}{tv} = \frac{s}{t}.$$

This follows by the following calculation, making use of 5.11:

$$\frac{sv}{tv} = (tv)^{-1}(sv) = v^{-1}t^{-1}sv = t^{-1}s = \frac{s}{t}.$$

As a generalization of 5.12, let s and u be arbitrary elements of F, and t and v arbitrary nonzero elements of F. Then we assert that

5.13 $$\frac{s}{t} = \frac{u}{v} \quad \text{if and only if} \quad sv = tu.$$

Suppose, first, that $s/t = u/v$, that is, that $t^{-1}s = v^{-1}u$. Multiplication by tv yields $sv = tu$. Conversely, if $sv = tu$, multiplication by $t^{-1}v^{-1}$ shows that $t^{-1}s = v^{-1}u$ or, otherwise expressed, that $s/t = u/v$.

The following are also easy to establish and will be left as exercises:

5.14

(i) $$\frac{s}{t} + \frac{u}{v} = \frac{sv + tu}{tv},$$

(ii) $$\frac{s}{t} \cdot \frac{u}{v} = \frac{su}{tv}.$$

Now a few remarks about exponents. If t is a nonzero element of F, we have a definition of t^{-1}; and if n is any positive integer, we now define t^{-n} to be $(t^{-1})^n$; also we define $t^0 = 1$. Under these definitions, the following laws of exponents hold for every choice of m and n as arbitrary integers (positive, negative, or zero), it being understood that t and v are arbitrary nonzero elements of F:

5.15

$$t^m \cdot t^n = t^{m+n},$$

$$\frac{t^m}{t^n} = t^{m-n},$$

$$(t^m)^n = t^{mn},$$

$$(tv)^m = t^m \cdot v^m,$$

$$\left(\frac{t}{v}\right)^m = \frac{t^m}{v^m}.$$

Of course, these are generalizations of the laws 2.27, 2.28, which hold for a commutative ring. Complete proofs of 5.15 can be given by mathematical induction.

EXERCISES

1. Let R be a ring with a finite number of elements, and r a nonzero element of R.
   (i) Show that there must exist a positive integer m such that $mr = 0$.
   (ii) Show that R cannot have characteristic zero.
2. Let a be a fixed nonzero element of an integral domain D such that $ma = 0$ for some positive integer m. Prove that the smallest such positive integer is the characteristic of D (that is, it is independent of the particular nonzero element a which is chosen). Show, by an example, that this result is not necessarily true for a *ring*.
3. Given the characteristics of rings R and S, what can you say about the characteristic of the direct sum $R \oplus S$?

4. Prove 5.14(i), (ii).

5. If s, t, u, and v are elements of a field F, prove (without using the laws of exponents) each of the following in which it is assumed that the necessary elements are different from zero:

(i) $$(t^{-1})^{-1} = t,$$

(ii) $$(-t)^{-1} = -(t^{-1}),$$

(iii) $$\left(\frac{s}{t}\right)^{-1} = \frac{t}{s},$$

(iv) $$\frac{\frac{s}{t}}{\frac{u}{v}} = \frac{vs}{ut},$$

(v) $$-\frac{s}{t} = \frac{(-s)}{t} = \frac{s}{(-t)},$$

(vi) $$\frac{s}{t} + \frac{u}{t} = \frac{s+u}{t},$$

(vii) $$\frac{s}{t} - \frac{u}{v} = \frac{sv - tu}{tv}.$$

6. Let c and d be distinct elements of a field F. If new operations of addition and multiplication are defined on F as follows:

$$x \oplus y = x + y - c, \quad x \odot y = c + \frac{(x-c)(y-c)}{d-c},$$

prove that one obtains a field F'. What are the zero and the unity of F'?

7. If F and F' are as in the preceding exercise, prove that the mapping $\theta: F \to F'$ defined by $x\theta = (d-c)x + c$, $x \in F$, is an isomorphism of F onto F'.

8. If a and b are elements of a commutative ring with characteristic the prime p, prove that $(a+b)^p = a^p + b^p$. Then generalize this result by proving that for every positive integer n, $(a+b)^{p^n} = a^{p^n} + b^{p^n}$.

5.4 THE FIELD OF RATIONAL NUMBERS

We now change our point of view as follows. Instead of studying properties of a given field, let us see how we can start with the integral domain **Z** of the integers and *construct* a field which contains **Z**. This is our first example of an important algebraic problem which may be stated in a general way as follows. Given an algebraic system U which does not have some specific property, to construct a larger system V which contains U and which does have the property in question. Naturally, this is not always possible, but it is in a number of interesting cases. At present, we start with the integral domain $\mathbf{Z}$ in which not every nonzero element has a multiplicative inverse, and shall construct a larger system—the field of rational numbers—which contains $\mathbf{Z}$ and in which every nonzero element necessarily has a multiplicative inverse. In this construction, our previous knowledge of the rational numbers will certainly be useful in suggesting procedure, but will be used in no other way.

Let S denote the set of all ordered pairs (a, b), where $a, b \in \mathbf{Z}$ and $b \neq 0$, that is,

$$S = \{(a, b) \mid a, b \in \mathbf{Z}, b \neq 0\}.$$

What we are going to do will be *suggested* by thinking of (a, b) as the familiar a/b, but we use an unfamiliar notation in order to clarify the logical procedure and to avoid using any property until we have actually proved it. If (a, b) and (c, d) are elements of S, we define $(a, b) \sim (c, d)$ to mean that $ad = bc$. Actually, "$\sim$" is an equivalence relation defined on S. The reflexive and symmetric properties are obviously true, and we now prove the transitive property. Suppose that $(a, b) \sim (c, d)$ and $(c, d) \sim (e, f)$, and let us show that $(a, b) \sim (e, f)$. Since $(a, b) \sim (c, d)$, we have $ad = bc$; and, similarly, we have $cf = de$. Multiplication of these equations by f and by b, respectively, yields $adf = bcf$ and $bcf = bde$. Hence $adf = bde$ and, since $d \neq 0$, it follows that $af = be$, that is, that $(a, b) \sim (e, f)$.

Now that we have an equivalence relation "$\sim$" defined on S, we follow a procedure somewhat like that previously used in obtaining the ring of integers modulo n. That is, we shall consider equivalence sets relative to "$\sim$," and give appropriate definitions of addition and multiplication of these sets.

If $(a, b) \in S$, according to our previous usage the equivalence set containing (a, b) would be designated by $[(a, b)]$. However, we shall now use the simpler notation $[a, b]$ to designate this equivalence set. In the sequel it is important to keep in mind that $[a, b] = [a_1, b_1]$ if and only if $(a, b) \sim (a_1, b_1)$, that is, if and only if $ab_1 = ba_1$. Of course, this is just the general property 1.11(i) of equivalence sets as applied in this particular case. The equivalence set $[a, b]$ may therefore be expressed as follows:

5.16 $$[a, b] = \{(x, y) \mid (x, y) \in S, xb = ya\}.$$

We now define addition and multiplication of equivalence sets as follows:

5.17 $$[a, b] + [c, d] = [ad + bc, bd]$$

and

5.18 $$[a, b] \cdot [c, d] = [ac, bd].$$

First, we observe that since (a, b) and (c, d) are elements of S, we have $b \neq 0$ and $d \neq 0$. Hence $bd \neq 0$, so that in fact $(ad + bc, bd)$ and (ac, bd) are elements of S and the right sides of 5.17 and 5.18 are equivalence sets.

Now, just as in the case of integers modulo n, we must show that addition and multiplication of equivalence sets are well-defined by 5.17 and 5.18. Suppose, then, that

5.19 $$[a, b] = [a_1, b_1] \quad \text{and} \quad [c, d] = [c_1, d_1].$$

In order to show that addition of equivalence sets is well-defined by 5.17 we must show that necessarily

$$[a, b] + [c, d] = [a_1, b_1] + [c_1, d_1],$$

that is, that

5.20 $$[ad + bc, bd] = [a_1d_1 + b_1c_1, b_1d_1].$$

From 5.19, we have that $ab_1 = ba_1$ and that $cd_1 = dc_1$. If we multiply the first of these equations by dd_1, the second by bb_1, and add the corresponding members, it follows that

$$(ad + bc)b_1d_1 = bd(a_1d_1 + b_1c_1).$$

However, this implies 5.20, and therefore addition of equivalence sets is well-defined by 5.17. The proof that multiplication is well-defined by 5.18 will be left as an exercise.

We may now state the following theorem.

5.21 Theorem. *Let* $\mathbf{Q}$ *denote the set of all equivalence sets of* S *relative to the equivalence relation "* $\sim$ *." Then with respect to the operations of addition and multiplication on* $\mathbf{Q}$ *defined by 5.17 and 5.18,* $\mathbf{Q}$ *is a field. Moreover, the set of all elements of* $\mathbf{Q}$ *of the form* $[a, 1]$, $a \in \mathbf{Z}$, *is a subring* $\mathbf{Z}'$ *of* $\mathbf{Q}$*; and the mapping* $\theta: \mathbf{Z} \to \mathbf{Z}'$, *defined by*

$$a\theta = [a, 1], \qquad a \in \mathbf{Z},$$

is an isomorphism of $\mathbf{Z}$ *onto* $\mathbf{Z}'$.

PROOF. The commutative laws of addition and multiplication, as well as the associative law of multiplication, are almost obvious, and we omit the details. The associative law of addition may be verified by the following straightforward calculation. Let $[a, b]$, $[c, d]$, and $[e, f]$ be elements of $\mathbf{Q}$. Then

$$\begin{aligned}([a, b] + [c, d]) + [e, f] &= [ad + bc, bd] + [e, f] \\ &= [adf + bcf + bde, bdf]\end{aligned}$$

and

$$\begin{aligned}[a, b] + ([c, d] + [e, f]) &= [a, b] + [cf + de, df] \\ &= [adf + bcf + bde, bdf],\end{aligned}$$

and we therefore have

$$([a, b] + [c, d]) + [e, f] = [a, b] + ([c, d] + [e, f]).$$

Since $[0, 1] + [a, b] = [a, b]$ and $[1, 1]\cdot[a, b] = [a, b]$, it follows that $[0, 1]$ is the zero and $[1, 1]$ the unity of $\mathbf{Q}$. However, if d is a nonzero integer, we have $[d, d] = [1, 1]$ and, similarly, $[0, 1] = [0, d]$. Hence the unity is $[d, d]$ and the zero is $[0, d]$ for *any* nonzero integer d. We may also observe that $[a, b] = [0, 1]$ if and only if $a = 0$, and to say, therefore, that $[a, b]$ is a nonzero element of $\mathbf{Q}$ is to say that $a \neq 0$.

Since $[a, b] + [-a, b] = [0, b^2]$ and $[0, b^2]$ is the zero of $\mathbf{Q}$, it follows that the additive inverse of $[a, b]$ is $[-a, b]$, that is, we have $-[a, b] = [-a, b]$, and each element of $\mathbf{Q}$ has an additive inverse.

One of the distributive laws is a consequence of the following calculations in which, at one point, we make use of the fact that $[b, b]$ is the unity of $\mathbf{Q}$:

$$\begin{aligned}[a, b]([c, d] + [e, f]) &= [a, b]\cdot[cf + de, df] \\ &= [acf + ade, bdf],\end{aligned}$$

$$\begin{aligned}[a, b]\cdot[c, d] + [a, b]\cdot[e, f] &= [ac, bd] + [ae, bf] \\ &= [acbf + bdae, b^2df] \\ &= [acf + ade, bdf]\cdot[b, b] \\ &= [acf + ade, bdf].\end{aligned}$$

The other distributive law is an immediate consequence of this one since multiplication is commutative.

Up to this point we have proved that $\mathbf{Q}$ is a commutative ring with unity. To prove that $\mathbf{Q}$ is a field, there remains only to show that every nonzero element of **Q** has a multiplicative inverse in **Q**. If $[a, b]$ is a nonzero element of **Q**, then $a \neq 0$ as well as $b \neq 0$, and it is clear that $[b, a] \in \mathbf{Q}$. Moreover,

$$[a, b] \cdot [b, a] = [ab, ab] = [1, 1],$$

and the multiplicative inverse of $[a, b]$ is $[b, a]$. That is, if $[a, b]$ is a nonzero element of **Q**, then $[a, b]^{-1} = [b, a]$. This completes the proof that **Q** is a field.

Now let $\mathbf{Z}'$ be the set of elements of **Q** of the form $[a, 1]$, $a \in \mathbf{Z}$, and consider the mapping $\theta \colon \mathbf{Z} \to \mathbf{Z}'$ defined by $a\theta = [a, 1]$, $a \in \mathbf{Z}$. This is clearly an onto mapping and it is also a one-one mapping since $[a, 1] = [b, 1]$ implies that $a = b$. Moreover, for $a, b \in \mathbf{Z}$, we have

$$(a + b)\theta = [a + b, 1] = [a, 1] + [b, 1] = a\theta + b\theta$$

and

$$(ab)\theta = [ab, 1] = [a, 1] \cdot [b, 1] = (a\theta)(b\theta).$$

Thus θ is an isomorphism of **Z** onto $\mathbf{Z}'$, and the theorem is established.

Since the subring $\mathbf{Z}'$ of **Q** is isomorphic to **Z**, we shall henceforth find it convenient to identify $\mathbf{Z}'$ with **Z** and, as a matter of notation, write simply a to designate the element $[a, 1]$ of **Q**. We may then consider that the field **Q** actually contains the ring **Z** of integers.

As a further simplification of notation, let us observe that

$$[a, b] = [a, 1] \cdot [1, b] = [a, 1] \cdot [b, 1]^{-1},$$

and hence we are justified in writing $a \cdot b^{-1}$ or a/b for the element $[a, b]$ of **Q**. Now that we have justified our familiar notation, we shall henceforth call an element of **Q** a *rational number* and the field **Q** the *field of rational numbers*. All of the notation of the preceding section naturally applies to the field **Q**. Throughout the rest of this book, **Q** will consistently be used to designate the field of rational numbers.

In the notation which we have finally introduced, the field **Q** consists of all numbers of the form a/b, where a and b are integers with $b \neq 0$, addition and multiplication being defined in the usual way (5.17, 5.18).

Let us emphasize the meaning of the notation we have introduced by considering, for example, the rational number 1/2. We are writing 1/2 for the equivalence set $[1, 2]$ used above. Now $[1, 2] =$

$[c, d]$ if and only if $d = 2c$, so we see that $1/2$ represents the equivalence set consisting of all ordered pairs of the form $(c, 2c)$, where c is a nonzero integer. Moreover, for example, $1/2 = 3/6$ simply because by our definition of equivalence, $(1, 2) \sim (3, 6)$ and therefore $[1, 2] = [3, 6]$.

Since $(-a)/b = a/(-b)$, we see that every rational number can be written in the form c/d, where $d > 0$. Moreover, if the integers c and d have a common nonzero factor k, so that $c = c_1k$ and $d = d_1k$, then $c/d = c_1/d_1$. It follows that every nonzero rational number r can be written uniquely in the form a/b, where a and b are relatively prime integers with $b > 0$. If r is expressed in this form, it is sometimes said that r is expressed *in lowest terms.*

Before proceeding to establish a few properties of the field $\mathbf{Q}$ of rational numbers, let us point out that in the above construction of the field $\mathbf{Q}$, the *only* properties of the integers which were used are those that imply that $\mathbf{Z}$ is an integral domain. Accordingly, by exactly the same construction we could start with an arbitrary integral domain D and obtain the *field of quotients* of D whose elements are expressible in the form ab^{-1}, where $a, b \in D$ with $b \neq 0$. In this terminology, the field $\mathbf{Q}$ of rational numbers is the field of quotients of $\mathbf{Z}$.

5.5 A FEW PROPERTIES OF THE FIELD OF RATIONAL NUMBERS

We have defined in Section 3.2 what we mean by an ordered integral domain. Since a field is necessarily an integral domain, by an *ordered field* we shall naturally mean a field which is an ordered integral domain. We shall now prove the following result.

5.22 Theorem. *Let $\mathbf{Q}^+$ denote the set of all rational numbers a/b, where a and b are integers such that $ab > 0$. Then $\mathbf{Q}^+$ has the properties 3.4 which define an ordered integral domain, and therefore the field $\mathbf{Q}$ is an ordered field whose positive elements are the elements of $\mathbf{Q}^+$.*

We may point out that when we write $ab > 0$, we mean that ab is a positive *integer* and we are only making use of the fact that $\mathbf{Z}$ is an ordered integral domain.

PROOF. First, we need to show that the definition of an element of $\mathbf{Q}^+$ does not depend upon the particular representation of a rational number. That is, we need to show that if $a/b = c/d$ and $ab > 0$, then also $cd > 0$. This follows from the observation that $a/b = c/d$ means that $ad = bc$ and $ab > 0$ implies that either a and b are both positive or they are both negative. The same must therefore be true of c and d; hence also $cd > 0$.

Now let us show (3.4(i)) that the set $\mathbf{Q}^+$ is closed under addition. Let a/b and c/d be elements of $\mathbf{Q}^+$, and therefore $ab > 0$ and $cd > 0$. Then

$$\frac{a}{b} + \frac{c}{d} = \frac{ad + bc}{bd},$$

and we wish to show that

$$(ad + bc)bd = abd^2 + cdb^2 > 0.$$

However, this inequality follows easily from the following known inequalities: $ab > 0$, $cd > 0$, $b^2 > 0$, and $d^2 > 0$.

It is trivial that $\mathbf{Q}^+$ is closed under multiplication (3.4(ii)). Moreover, if a/b is a nonzero rational number, then either $ab > 0$ or $ab < 0$. It follows that for every rational number a/b, exactly one of the following holds (3.4(iii)):

$$\frac{a}{b} = 0, \quad \frac{a}{b} > 0, \quad -\frac{a}{b} > 0.$$

Hence $\mathbf{Q}^+$ has the three required properties, and the field $\mathbf{Q}$ of rational numbers is ordered.

It will be observed that what we have done is to make use of the known ordering of the integers to establish an ordering of the rational numbers. Inasmuch as we have identified the integer a with the rational number $a/1$, it is clear that a is a positive integer if and only if a is a positive rational number. In other words, our ordering of the rational numbers is an *extension* of the previous ordering of the integers.

In view of Theorem 5.22, we can introduce inequalities involving rational numbers in the usual way. That is, if $r, s \in \mathbf{Q}$, we write $r > s$ (or $s < r$) to mean that $r - s \in \mathbf{Q}^+$, and so on. We now have available all the usual properties (3.6) of inequalities for rational numbers. In the future we shall make use of these properties without specific reference.

The following is a significant property of the rational numbers.

5.23 Theorem. *Between any two distinct rational numbers there is another rational number.*

PROOF. Suppose that $r, s \in \mathbf{Q}$ with $r < s$. The theorem will be established by showing that

$$r < \frac{r+s}{2} < s,$$

and hence that $(r + s)/2$ is a rational number between r and s. Since $r < s$, we have $r + r < r + s$, or $2r < r + s$. Now multiplying this last inequality by the positive rational number $1/2$, we obtain $r < (r + s)/2$. In a similar manner, it can be shown that $(r + s)/2 < s$, and we omit the details.

The property of the rational numbers stated in the preceding theorem is often expressed by saying that the rational numbers are *dense*. We shall now prove in the following theorem another simple, but important, property of the rational numbers.

5.24 Theorem (ARCHIMEDEAN PROPERTY). *If r and s are any positive rational numbers, there exists a positive integer n such that $nr > s$.*

PROOF. Let $r = a/b$, $s = c/d$, where a, b, c, and d are positive integers. If n is a positive integer, then $n(a/b) > c/d$ if and only if $n(ad) > bc$. We now assert that this last inequality is necessarily satisfied if we choose $n = 2bc$. For $ad \geq 1$, and therefore $2ad > 1$. Multiplying this inequality by the positive integer bc shows that $2adbc > bc$. Hence, $n = 2bc$ certainly satisfies our requirement. Of course, we do not mean to imply that this is necessarily the smallest possible choice of n.

5.6 SUBFIELDS AND EXTENSIONS

Let us make the following convenient definition.

5.25 Definition. A subring F' of a field F which is itself a field is called a *subfield* of F. If F' is a subfield of F, F is frequently called an *extension* of F'.

Although we are here primarily interested in fields, we shall first prove the following fairly general result (cf. Exercise 14 of Section 2.5).

5.26 Theorem. *Let D be an integral domain with unity e, and suppose that R is a subring of D having more than one element. Then, if R has a unity, $e \in R$ and e is the unity of R.*

PROOF. Let f be the unity of R and let us prove that $f = e$. Clearly,

$$f(fe - e) = f^2e - fe = fe - fe = 0.$$

However, $f \neq 0$ since R has more than one element, and therefore f is not a divisor of zero in the integral domain D. Hence we must have $fe - e = 0$, or $fe = e$. But e is the unity of D and $f \in D$, so $fe = f$ and we conclude that $f = e$, as we wished to show.

Since a field F is an integral domain and a subfield of F necessarily has a unity and has more than one element, we have the following special case of the result just obtained.

5.27 Corollary. *The unity of a field F is also the unity of each subfield of F.*

We may point out that this corollary and Theorem 5.8 show that a field and all of its subfields have the same characteristic.

Since $\mathbf{Z}_p$ is a field for each prime p, the first statement of the following theorem has already been established as a part of Theorem 5.10.

5.28 Theorem. *A field of characteristic p contains a subfield which is isomorphic to the field $\mathbf{Z}_p$. A field of characteristic zero contains a subfield which is isomorphic to the field $\mathbf{Q}$.*

PROOF. To prove the second statement of the theorem, let F be a field of characteristic zero and with e as unity. We already know by Theorem 5.10 that the subring $\mathbf{Z}e$ of F is isomorphic to $\mathbf{Z}$. It is perhaps then not surprising that F contains a subfield which is isomorphic to the field $\mathbf{Q}$ of quotients of $\mathbf{Z}$. However, we shall present some of the steps in the proof of this fact.

Since F has characteristic zero, if n is a nonzero integer, then $ne \neq 0$ and therefore ne has a multiplicative inverse $(ne)^{-1}$ in F. Let $\mathbf{Q}'$ be the set of all elements of F expressible in the form $(me)(ne)^{-1}$, where $m, n \in \mathbf{Z}$ with $n \neq 0$. We leave it to the reader to verify that $\mathbf{Q}'$ is a subfield of F. (Of course, it is the field of quotients of the integral domain $\mathbf{Z}e$.) We proceed to show that there exists an isomorphism of **Q** onto **Q**′. To this end, we start by defining a mapping $\alpha: \mathbf{Q} \to \mathbf{Q}'$ as follows:

$$(mn^{-1})\alpha = (me)(ne)^{-1}, \qquad m, n \in \mathbf{Z}, n \neq 0.$$

Since elements of **Q** are not uniquely expressible in the form mn^{-1}, we must first show that α is in fact a well-defined mapping of **Q** into **Q**′. Suppose that m, n, k, and l are integers with $n \neq 0$ and $l \neq 0$ such that in **Q**, $mn^{-1} = kl^{-1}$. Then $ml = nk$ and therefore in **Q**′, $(me)(le) = (ne)(ke)$. It follows that $(me)(ne)^{-1} = (ke)(le)^{-1}$ and the mapping α is indeed well-defined. To verify that it is a one-one mapping, suppose that $(mn^{-1})\alpha = (kl^{-1})\alpha$, that is, that $(me)(ne)^{-1} = (ke)(le)^{-1}$. From this equation it follows that $(ml - nk)e = 0$. Then, since F has characteristic zero, we must have $ml - nk = 0$ or $mn^{-1} = kl^{-1}$. Thus α is a one-one mapping. It is clearly an onto mapping and we leave as an exercise the verification that the operations of addition and multiplication are preserved, and hence that α is the desired isomorphism.

In view of this theorem, it is clear that the only fields which do not have proper subfields are **Q** and the fields $\mathbf{Z}_p$. Moreover, if we do not consider isomorphic fields as "different," every field is an extension of the rational field **Q** or of one of the fields $\mathbf{Z}_p$ for some prime p. The study of extensions of a given field is an important part of the general theory of fields, but we shall not give a systematic account of this topic in this book. However, in a later chapter we shall present a few illustrations of one important method of constructing extensions of a given field.

EXERCISES

1. Prove that multiplication of equivalence sets is well-defined by 5.18.
2. Go through the proof of Theorem 5.21 and verify that all the steps can be carried out with an arbitrary integral domain D in place of **Z**, thus obtaining the field of quotients of D. In this construction, why cannot an arbitrary commutative ring be used in place of **Z**?
3. Complete the proof of Theorem 5.23, and state what properties of inequalities have been used.
4. If u and v are positive rational numbers with $u < v$, show that $1/u > 1/v$.

5. If $r, s \in \mathbf{Q}$ and $r < s$, and $u, v \in \mathbf{Q}^+$, show that

$$r < \frac{ur + vs}{u + v} < s.$$

6. If $r, s \in \mathbf{Q}$ with $r < s$, and n is an arbitrary positive integer, show that there exist rational numbers $t_1, t_2, \cdots, t_n$ such that

$$r < t_1 < t_2 < \cdots < t_n < s.$$

7. Prove that addition and multiplication are preserved under the mapping α defined in the proof of Theorem 5.28.
8. Prove that a subring R (with more than one element) of a field F is a subfield of F if and only if the multiplicative inverse in F of each nonzero element of R is an element of R.
9. Let D and D' be integral domains with respective fields of quotients F and F'. If $\theta: D \to D'$ is an isomorphism of D onto D', prove that the mapping $\alpha: F \to F'$ defined by

$$(ab^{-1})\alpha = (a\theta)(b\theta)^{-1}, \qquad a, b \in D, b \neq 0$$

is well-defined and is an isomorphism of F onto F'.

★ 5.7 CONSTRUCTION OF THE INTEGERS FROM THE NATURAL NUMBERS

We indicated in Section 3.5 how all the familiar properties of the natural numbers, that is, the positive integers, can be obtained from a few simple axioms. Let N be the system of all natural numbers. In this system we have operations of addition and multiplication, and all the properties of an integral domain hold except that there is no zero, and elements do not have additive inverses. In this section we shall outline a procedure by which we can start with N and *construct* the ring $\mathbf{Z}$ of all integers. The method closely parallels that by which we have constructed the rational numbers from the ring of integers. We may emphasize that we now assume as known only the properties of the natural numbers.

Let T be the set of all ordered pairs (a, b) of elements of N. Our procedure will be *suggested* by thinking of (a, b) as meaning $a - b$, but we must so formulate our statements that only natural numbers are involved. If (a, b) and (c, d) are elements of T, we shall write $(a, b) \sim (c, d)$ to mean that $a + d = b + c$. It is easy to verify that

"$\sim$" is an equivalence relation on T. One way to characterize the equivalence set $[a, b]$ which contains (a, b) is as follows:

$$[a, b] = \{(x, y) \mid x, y \in N, x + b = y + a\}.$$

Now let $\mathbf{Z}$ be the set of all such equivalence sets, and let us make the following definitions:

5.29 $$[a, b] + [c, d] = [a + c, b + d],$$

5.30 $$[a, b] \cdot [c, d] = [ac + bd, ad + bc].$$

It can be shown that addition and multiplication are well-defined, and hence that we have operations of addition and multiplication defined on $\mathbf{Z}$. The following theorem can now be established.

5.31 Theorem. *With respect to the definitions 5.29 and 5.30 of addition and multiplication,* $\mathbf{Z}$ *is an integral domain and, by a suitable change of notation, we may consider that* $\mathbf{Z}$ *contains the set* N *of natural numbers. If we now define the set* $\mathbf{Z}^+$ *of positive elements of* $\mathbf{Z}$ *to be the set* N*, then* $\mathbf{Z}$ *is an ordered integral domain in which the set of positive elements is well-ordered.*

We shall make a few remarks about the proof of this theorem, but shall not write out all the details. The zero of $\mathbf{Z}$ is $[c, c]$ for an arbitrary natural number c. The additive inverse of $[a, b]$ is $[b, a]$, that is, $-[a, b] = [b, a]$.

PROOF. Let N' be the set of all elements of $\mathbf{Z}$ of the form $[x + 1, 1]$, $x \in N$. Then the mapping $\theta: N \to N'$, defined by

$$x\theta = [x + 1, 1], \qquad x \in N,$$

is a one-one mapping of N onto N' and, moreover, addition and multiplication are preserved under this mapping. Hence, as a matter of notation, let us identify N' with N; that is, let us write x in place of $[x + 1, 1]$ so that $\mathbf{Z}$ now actually contains N. If $[a, b] \in \mathbf{Z}$ it is easy to verify that

$$\begin{aligned}[a, b] &= [a + 1, 1] + [1, b + 1] \\ &= [a + 1, 1] - [b + 1, 1] \\ &= a - b.\end{aligned}$$

We have therefore justified writing $a - b$ in place of $[a, b]$.

If $c, d \in N$, we defined $c > d$ in Section 3.5 to mean that there exists a natural number e such that $c = d + e$. Since $[a, b] = a - b$, we see that $[a, b]$ is an element of N if $a > b$, and that $-[a, b]$ is an element of N if $b > a$. The elements of **Z** therefore consist of the natural numbers, the additive inverses of the natural numbers, and zero. Of course, the integral domain **Z** is called the *ring of integers*.

If we set $\mathbf{Z}^+ = N$, then $\mathbf{Z}^+$ has the properties (3.4) which make **Z** an ordered integral domain. Finally, then, since the set N is well-ordered, we have that **Z** is an ordered integral domain in which the set of positive elements is well-ordered. Our viewpoint in this book has been to *assume* that the ring of integers has all the properties implied in this statement. However, we have now indicated how this result can be proved by starting only with the Peano Axioms for the natural numbers.

NOTES AND REFERENCES

Many texts on abstract algebra contain fairly extensive discussions of the theory of fields. In particular, see items [1], [2], [3], [5], [6], [8], and [13] of the Bibliography. Look in the indices for *fields*, *field extensions*, or *Galois Theory*.

Chapter 6

Real and Complex Numbers

In the preceding chapter we gave a detailed construction of the field of rational numbers, starting with the integral domain of integers. In this chapter we shall be concerned with extensions of the field of rational numbers to the field of real numbers and of the field of real numbers to the field of complex numbers. There are different methods of carrying out the first of these extensions but any one of them involves rather long and detailed calculations. Accordingly, instead of presenting the details, we shall merely state the existence of a certain extension of the field of rational numbers which we shall call the field of real numbers, and briefly discuss a few properties of this field. It is quite easy to construct the field of complex numbers from the field of real numbers and we shall carry out this construction and establish a number of fundamental properties of the complex numbers.

6.1 THE FIELD OF REAL NUMBERS

The rational numbers are sufficient for use in all simple applications of mathematics to physical problems. For example, measurements are usually given to a certain number of decimal places, and any finite decimal is a rational number. However, from a theoretical point of view, the system of rational numbers is entirely inadequate. The Pythagoreans made this discovery about 500 B.C. and were profoundly shocked by it. Consider, for example, an isosceles right triangle whose legs are 1 unit in length. Then, by the Pythagorean theorem, the hypotenuse has length $\sqrt{2}$; and from this geometrical consideration it appears that there must exist a "number" $\sqrt{2}$, although we have shown in Section 4.5 that it cannot be a rational number.

The inherent difficulty in extending the field of rational numbers to the field of real numbers is perhaps indicated by the fact that a satisfactory theory of the real numbers was not obtained until the latter half of the nineteenth century. Although other men also made contributions to the theory, it is usually attributed to the German mathematicians Dedekind (1831–1916) and Cantor (1845–1918). We shall not present here the work of either of these men but shall presently state without proof the fundamental theorem which each of them essentially proved and by quite different methods.* In order to do this, we must first make a few preliminary definitions.

So far, the only ordered field which we have studied is the field $\mathbf{Q}$ of rational numbers. However, for the moment, suppose that F is an arbitrary ordered field and let us make the following definition.

6.1 Definition. Let S be a set of elements of an ordered field F. If there exists an element b of F such that $x \leq b$ for every element x of S, then b is called an *upper bound* of the set S in F.

As an example, the set $S_1 = \{\frac{1}{2}, 1, 2\}$ of elements of $\mathbf{Q}$ has an upper bound 2. Also $\frac{5}{2}$ is an upper bound of this set, as is 117, and so on. Thus, if a set has an upper bound, it has many upper bounds. Clearly, the set $\mathbf{Z}^+$ of all positive integers does not have an upper bound in $\mathbf{Q}$. As another example, consider the set

$$S_2 = \{a \mid a \in \mathbf{Q}, a > 0, a^2 < 2\}.$$

Then S_2 has upper bounds in $\mathbf{Q}$, one of them being 3.

6.2 Definition. Let S be a set of elements of an ordered field F. If there exists an upper bound c of S in F such that no smaller element of F is an upper bound of S, then c is called the *least upper bound* (l.u.b.) of S in F.

It follows from this definition that if a set S has a l.u.b., it is unique. Moreover, if c is the l.u.b. of the set S in F and $d \in F$ such that $d < c$, then there must exist an element s of S such that $s > d$ since,

* See the notes and references at the end of this chapter.

otherwise, d would be an upper bound of S less than the least upper bound.

For the set S_1 exhibited above, the element 2 of S_1 is clearly the l.u.b. of S_1 in $\mathbf{Q}$. However, for the set S_2 the situation is not quite so obvious. Although we shall not give the details, it is true and should perhaps not be surprising that there exists no rational number which is the l.u.b. of the set S_2. That is, if $c \in \mathbf{Q}$ is an upper bound of S_2, there exists $d \in \mathbf{Q}$ such that $d < c$ is also an upper bound of S_2. Therefore S_2 has no l.u.b. in $\mathbf{Q}$. Thus we have an example of a set of elements of $\mathbf{Q}$ which has upper bounds in $\mathbf{Q}$ but no l.u.b. in $\mathbf{Q}$. In the field of real numbers, whose existence is asserted in the next theorem, this situation cannot arise. In fact, it is the existence of an ordered field with this property which may be considered to be the principal contribution of Dedekind and Cantor to the subject. Let us state this result as the following theorem.

6.3 Theorem. *There exists a field* $\mathbf{R}$, *called the field of real numbers, with the following properties:*

(*i*) $\mathbf{R}$ *is an extension of the field* $\mathbf{Q}$ *of rational numbers. Moreover,* $\mathbf{R}$ *is an ordered field and* $\mathbf{Q}^+ \subset \mathbf{R}^+$.

(*ii*) *If* S *is a nonempty set of elements of* $\mathbf{R}$ *which has an upper bound in* $\mathbf{R}$, *it has a l.u.b. in* $\mathbf{R}$.

The elements of $\mathbf{R}$ are called *real numbers*. An element of $\mathbf{R}$ which is not an element of $\mathbf{Q}$ is called an *irrational* number. Considered as elements of $\mathbf{R}$, the set S_2 defined above has a l.u.b. (by 6.3(ii)) and this l.u.b. we may *define* to be the number $\sqrt{2}$. Clearly, $\sqrt{2}$ is an irrational number and it is the fact that this number is not an element of $\mathbf{Q}$ which prevents S_2 from having a l.u.b. *in the field* $\mathbf{Q}$.

The fact that $\mathbf{Q}^+ \subset \mathbf{R}^+$ is sometimes expressed by saying that the ordering of $\mathbf{R}$ is an extension of the ordering of $\mathbf{Q}$. That is, a rational number is a positive rational number if and only if, considered as a real number, it is a positive real number. A similar situation arose when we passed from the integers to the rational numbers.

6.2 SOME PROPERTIES OF THE FIELD OF REAL NUMBERS

In this section we shall prove two fundamental properties of real numbers and state one additional property without proof. Of course, our proofs will be based on the assumed properties (i) and (ii) of Theorem 6.3.

Throughout the rest of this book we shall continue to denote the field of real numbers by $\mathbf{R}$ and the set of positive real numbers by $\mathbf{R}^+$.

6.4 Theorem (Archimedean Property). *If a, $b \in \mathbf{R}^+$, there exists a positive integer n such that $na > b$.*

PROOF. Let us assume that $ka \leq b$ for every positive integer k, and seek a contradiction. Another way of stating this assumption is to asssert that b is an upper bound of the set $S = \{ka \mid k \in \mathbf{Z}^+\}$. Since this set has an upper bound, by 6.3(ii) it has a l.u.b., say c. Now $c - a < c$ and therefore $c - a$ is not an upper bound of the set S. This implies that there exists an element la of S, $l \in \mathbf{Z}^+$, such that $la > c - a$. It follows that $(l + 1)a > c$ and since $(l + 1)a \in S$, we have a contradiction of the fact that c is the l.u.b. of the set S. The proof is therefore complete.

It was shown in Section 5.5 that between any two distinct rational numbers there is another rational number. A generalization of the result is given in the following theorem.

6.5 Theorem. *If a, $b \in \mathbf{R}$ with $a < b$, there exists a rational number m/n ($m, n \in \mathbf{Z}$) such that*

$$a < \frac{m}{n} < b.$$

PROOF. For simplicity, we shall assume that $a > 0$ and leave the rest of the proof as an exercise.

Since $b - a > 0$, by the preceding theorem there exists $n \in \mathbf{Z}^+$ such that $n(b - a) > 1$. Let n be some such fixed integer. Again applying the preceding theorem to the real numbers 1 and na, there exists $m \in \mathbf{Z}^+$ such that $m > na$, and let m be the *least* positive integer with this property. Now $m > na$ implies that $a < m/n$ and we proceed to complete the proof by showing that also $m/n < b$ or, equivalently, that $m < nb$. Suppose that $m \geq nb$. Since $n(b - a) > 1$, we have $m \geq nb > na + 1$. Thus $m > 1$ and $(m - 1) \in \mathbf{Z}^+$ such that $(m - 1) > na$. Since $m - 1 < m$, this violates our choice of m as the least positive integer which is greater than na. Our assumption that $m \geq nb$ has led to a contradiction, and we conclude that $m < nb$. This completes the proof.

In particular, this theorem tells us that between any two irrational numbers there is a rational number. It is also true that between any two rational numbers there is an irrational number. (See Exercise 2 below.) Thus the rational and irrational numbers are very closely intertwined.

Although it is true that all the properties of the real numbers can be established using only the properties (i) and (ii) of Theorem 6.3, we shall give no further proofs in this book. However, let us conclude this brief discussion of the real numbers by stating without proof the following familiar and important result.

6.6 Theorem. *For each positive real number a and each positive integer n, there exists exactly one positive real number x such that $x^n = a$.*

The real number x whose existence is asserted by this theorem may be called the *principal* nth root of a and designated by the familiar notation $a^{1/n}$ or by $\sqrt[n]{a}$.

EXERCISES

1. Define lower bound and greatest lower bound of a set of elements of an ordered field. Prove that if a nonempty set of elements of **R** has a lower bound, it has a greatest lower bound in **R**. [Hint: Consider the additive inverses of elements of the set.]
2. Prove that if $a, b \in \mathbf{R}$ with $a < b$, then
$$a < a + \frac{b - a}{\sqrt{2}} < b.$$
Hence prove that between any two distinct rational numbers there is an irrational number.
3. Complete the proof of Theorem 6.5 by showing that the stated result holds also for the case in which $a \leq 0$.
4. Let S_1 and S_2 be nonempty sets of real numbers having, respectively, b_1 and b_2 as least upper bounds. If $S_3 = \{s_1 + s_2 \mid s_1 \in S_1, s_2 \in S_2\}$, prove that $b_1 + b_2$ is the least upper bound of the set S_3.

6.3 THE FIELD OF COMPLEX NUMBERS

In order to construct the field of complex numbers, we begin by considering ordered pairs (a, b) of *real* numbers. Our definitions of addition and multiplication will be motivated by the formal properties of expressions of the form $a + bi$, where $i^2 = -1$. However, we are not justified in assuming that there *is* a "number" whose square is -1 until we have constructed a field which has an element with this property. Accordingly, as in the case of the construction of the rational numbers, we begin with an unfamiliar notation in order to avoid using any property until we have established it. We may remind the reader that the equal sign is being used in the sense of identity, that is, $(a, b) = (c, d)$ means that $a = c$ and $b = d$.

We proceed to prove the following theorem, which establishes the existence of the field we shall presently call the field of complex numbers.

6.7 Theorem. *Let* $\mathbf{C}$ *be the set of all ordered pairs* (a, b) *of elements of the field* $\mathbf{R}$ *of real numbers, and let us define operations of addition and multiplication on* $\mathbf{C}$ *as follows:*

6.8 $$(a, b) + (c, d) = (a + c, b + d),$$

6.9 $$(a, b)(c, d) = (ac - bd, ad + bc).$$

Then $\mathbf{C}$ *is a field with respect to these definitions of addition and multiplication. Moreover, the set of all elements of* $\mathbf{C}$ *of the form* $(a, 0)$, $a \in \mathbf{R}$, *is a subfield of* $\mathbf{C}$ *which is isomorphic to the field* $\mathbf{R}$.

PROOF. The required properties of addition are almost obvious. From 6.8, it follows that addition is commutative and associative, that $(0, 0)$ is the zero of $\mathbf{C}$, and that the additive inverse of (a, b) is $(-a, -b)$.

The associative law of multiplication is a consequence of the following straightforward calculations:

$$\begin{aligned}((a, b)(c, d))(e, f) &\\ &= (ac - bd, ad + bc)(e, f)\\ &= (ace - bde - adf - bcf, acf - bdf + ade + bce),\end{aligned}$$

$$\begin{aligned}(a, b)((c, d)(e, f)) &\\ &= (a, b)(ce - df, cf + de)\\ &= (ace - adf - bcf - bde, acf + ade + bce - bdf),\end{aligned}$$

and these turn out to be equal elements of $\mathbf{C}$.

Next, let us verify one of the distributive laws as follows:

$$\begin{aligned}(a, b)((c, d) + (e, f)) &\\ &= (a, b)(c + e, d + f)\\ &= (ac + ae - bd - bf, ad + af + bc + be),\end{aligned}$$

$$\begin{aligned}(a, b)(c, d) + (a, b)(e, f) &\\ &= (ac - bd, ad + bc) + (ae - bf, af + be)\\ &= (ac - bd + ae - bf, ad + bc + af + be),\end{aligned}$$

and again we have equal elements of **C**. The other distributive law follows from this one as soon as we show that multiplication is commutative, and the commutativity of multiplication follows easily from 6.9. For, by interchanging (a, b) and (c, d) in 6.9, we see that

$$(c, d)(a, b) = (ca - db, cb + da),$$

and the right side of this equation is equal to the right side of 6.9. Hence,

$$(a, b)(c, d) = (c, d)(a, b).$$

We have now proved that **C** is a commutative ring, and it is easily verified that it has the unity $(1, 0)$. To show that **C** is a field, we need only show that each nonzero element (a, b) of **C** has a multiplicative inverse in **C**. Since the zero is $(0, 0)$, to say that (a, b) is not the zero of **C** is to say that a and b are not both equal to zero. Since a is an element of the ordered field **R**, we know that if $a \neq 0$, then $a^2 > 0$. Similarly, if $b \neq 0$, we have $b^2 > 0$. It follows that if (a, b) is not the zero of **C**, then necessarily $a^2 + b^2 > 0$ and, in particular, $a^2 + b^2 \neq 0$. Hence,

$$\left(\frac{a}{a^2 + b^2}, \frac{-b}{a^2 + b^2}\right)$$

is an element of **C** and it may be verified by direct calculation (using 6.9) that

$$(a, b)\left(\frac{a}{a^2 + b^2}, \frac{-b}{a^2 + b^2}\right) = (1, 0).$$

We have therefore shown that every nonzero element of **C** has a multiplicative inverse in **C**, and hence we have proved that **C** is a field.

To complete the proof of the theorem, let $\mathbf{R}'$ be the set of all elements of **C** of the form $(a, 0)$, $a \in \mathbf{R}$. Then the mapping $\theta: \mathbf{R}' \to \mathbf{R}$ defined by $(a, 0)\theta = a$, $a \in \mathbf{R}$, is a one-one mapping of $\mathbf{R}'$ onto **R**. Moreover,

$$[(a, 0) + (b, 0)]\theta = (a + b, 0)\theta = a + b = (a, 0)\theta + (b, 0)\theta$$

and

$$[(a, 0)(b, 0)]\theta = (ab, 0)\theta = ab = [(a, 0)\theta][(b, 0)\theta].$$

Hence, the operations of addition and multiplication are preserved under this mapping, and the mapping therefore defines an isomorphism of $\mathbf{R}'$ onto $\mathbf{R}$. This completes the proof of the theorem.

An element of the field which we have constructed is called a *complex number*, and $\mathbf{C}$ is called the *field of complex numbers*.

We shall henceforth adopt a more familiar notation by identifying $\mathbf{R}'$ with $\mathbf{R}$, that is, we shall write a in place of $(a, 0)$, and consider that the field $\mathbf{C}$ of complex numbers actually contains the field $\mathbf{R}$ of real numbers. Also, for simplicity of notation, as well as for historical reasons, we shall use the symbol i to designate the particular element $(0, 1)$ of $\mathbf{C}$. Since $(0, 1)^2 = (-1, 0)$, in our new notation we have $i^2 = -1$. Now it is easily verified that

$$(a, 0) + (b, 0)(0, 1) = (a, b)$$

and, using the notation we have introduced, it follows that $a + bi = (a, b)$. Accordingly, in the future we shall write $a + bi$ in place of (a, b). In this notation, the product 6.9 of two elements of $\mathbf{C}$ may be expressed in the following form:

6.10 $$(a + bi)(c + di) = ac - bd + (ad + bc)i.$$

Of course, the right side of 6.10 may be obtained from the left by multiplying out with the aid of the usual distributive, associative, and commutative laws, and replacing i^2 by -1.

We have now extended the field of real numbers to the field of complex numbers. It should be pointed out, however, that one familiar property of the field of rational numbers and of the field of real numbers does not carry over to the field of complex numbers.

6.11 Theorem. *The field* $\mathbf{C}$ *of complex numbers is not an ordered field.*

PROOF. By this statement we mean that there does not exist any set $\mathbf{C}^+$ of elements of $\mathbf{C}$ having the properties (3.4) required for $\mathbf{C}$ to be an ordered field. This fact is a consequence of the following observations. If $\mathbf{C}$ were ordered, 3.6(v) would show that the square of every nonzero element would be positive; in particular, both i^2 and 1 would be positive. Then -1 would be negative, and we have a contradiction since $i^2 = -1$.

The fact that $\mathbf{C}$ is not ordered means that inequalities cannot be used between complex numbers. In other words, it is meaningless to speak of one complex number as being greater or less than another.

Throughout the rest of this book we shall continue to denote the field of complex numbers by **C**.

6.4 THE CONJUGATE OF A COMPLEX NUMBER

Let us make the following definition.

6.12 Definition. If $u = a + bi \in \mathbf{C}$, we define the *conjugate* of u to be the element u^* of **C** given by: $u^* = a - bi$.†

As examples, we have $(1 + 7i)^* = 1 - 7i$, $(2 - 2i)^* = 2 + 2i$, $4^* = 4$, and so on.

Now the mapping $\alpha: \mathbf{C} \to \mathbf{C}$ defined by $u\alpha = u^*$, $u \in \mathbf{C}$, is a one-one mapping of **C** onto **C**. We proceed to show that the operations of addition and multiplication are preserved under this mapping. Let $u = a + bi$ and $v = c + di$ be elements of **C**. Then

$$\begin{aligned}(u + v)\alpha = (u + v)^* &= [(a + c) + (b + d)i]^* = a + c - (b + d)i \\ &= (a - bi) + (c - di) = u^* + v^* = u\alpha + v\alpha\end{aligned}$$

and

$$\begin{aligned}(uv)\alpha = (uv)^* &= [ac - bd + (ad + bc)i]^* = ac - bd - (ad + bc)i \\ &= (a - bi)(c - di) = u^*v^* = (u\alpha)(v\alpha).\end{aligned}$$

Since the operations of addition and multiplication are preserved under the mapping α, it follows that this is an ismorphism of the field **C** onto itself. Such an isomorphism is frequently called an *automorphism.*

In working with complex numbers, the concept of conjugate plays an important role. A number of simple, but significant, properties are presented in Exercise 2 which follows.

EXERCISES

1. Find the multiplicative inverse of the nonzero element (a, b) of **C** by assuming that r and s are real numbers such that $(a, b)(r, s) = (1, 0)$, and solving for r and s.

† Historically, the usual notation for the conjugate of a complex number z is $\bar{z}$ instead of z^*. The present notation has been adopted only because it makes it easier to print such expressions as the conjugate of the sum of two or more complex numbers.

2. Prove each of the following:
   (i) If $u \in \mathbf{C}$, then $uu^* \in \mathbf{R}$ and $u + u^* \in \mathbf{R}$; moreover, if $u \neq 0$, then $uu^* > 0$.
   (ii) If $u \in \mathbf{C}$, then $(u^*)^* = u$.
   (iii) If $u \in \mathbf{C}$ and $u \neq 0$, then $(u^{-1})^* = (u^*)^{-1}$.
   (iv) If $u \in \mathbf{C}$, then $u = u^*$ if and only if $u \in \mathbf{R}$.
   (v) If $u \in \mathbf{C}$ and n is a positive integer, then $(u^n)^* = (u^*)^n$.
3. If $\alpha: \mathbf{C} \rightarrow \mathbf{C}$ is the isomorphism defined above, part (iv) of the preceding exercise shows that $u\alpha = u$ if and only if $u \in \mathbf{R}$. Prove that if $\phi: \mathbf{C} \rightarrow \mathbf{C}$ is an isomorphism of $\mathbf{C}$ onto $\mathbf{C}$ with the property that $a\phi = a$ for every $a \in \mathbf{R}$, then $\phi = \alpha$ or ϕ is the identity mapping in the sense that *every* element of $\mathbf{C}$ is its own image. [Hint: Consider the possibilities for $i\phi$.]
4. Let S be an arbitrary ring and T the set of all ordered pairs (a, b) of elements of S. If addition and multiplication are defined by 6.8 and 6.9, respectively, verify each of the following:
   (i) T is a ring,
   (ii) T is a commutative ring if and only if S is a commutative ring,
   (iii) T has a unity if and only if S has a unity.
5. In the notation of the preceding exercise, let S be the ring $\mathbf{Z}_2$ of integers modulo 2. Exhibit addition and multiplication tables for the corresponding ring T. Is T a field in this case? Is it an integral domain? Is it isomorphic to any of the rings with four elements given in Chapter 2?

6.5 GEOMETRIC REPRESENTATION AND TRIGONOMETRIC FORM

It is implicit in our construction of the complex numbers that the mapping $a + bi \rightarrow (a, b)$ is a one-one mapping of the set $\mathbf{C}$ of all complex numbers onto the set of all ordered pairs of real numbers. Now in ordinary plane analytic geometry we represent points in the plane by their coordinates, that is, by ordered pairs of real numbers. Accordingly, we may represent a point in the plane by a single complex number. In other words, we shall sometimes find it convenient to associate with the complex number $a + bi$ the point with rectangular coordinates (a, b), and to say that this point has *coordinate* $a + bi$. A number of examples are given in Figure 9. It will be observed that a real number, that is, a complex number of the form $a + 0i$, is the coordinate of a point on the x-axis. A number of the form $0 + bi$, sometimes called a *pure imaginary*, is the coordinate of a point on the y-axis. We may also observe that a complex number $a + bi$ and its conjugate $a - bi$ are coordinates of points that are symmetrically located with respect to the x-axis.

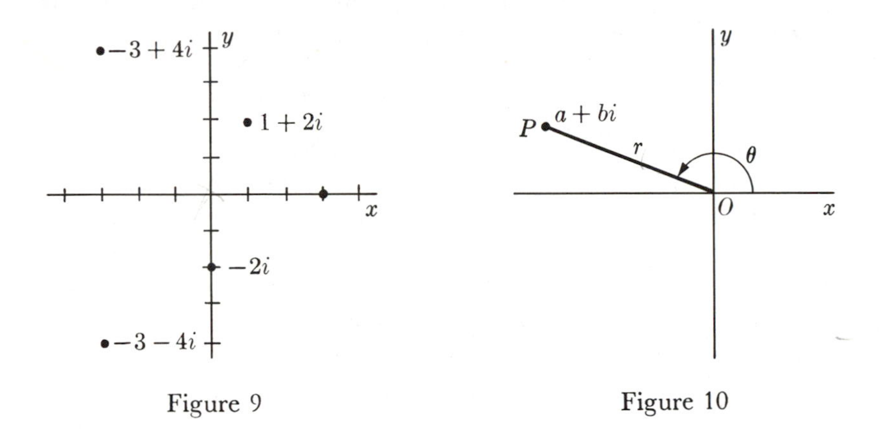

Figure 9

Figure 10

Instead of specifying points in a plane by means of rectangular coordinates, we may of course use polar coordinates. If P is the point with nonzero coordinate $a + bi$, the distance of P from the origin O of coordinates is the positive real number $r = \sqrt{a^2 + b^2}$. If θ is an angle in standard position with terminal side OP, as in Figure 10, then by the definition of the trigonometric functions we have

$$a = r\cos\theta, \quad b = r\sin\theta.$$

It follows that the complex number $a + bi$ can be expressed in the form

6.13 $$a + bi = r(\cos\theta + i\sin\theta).$$

We have been assuming that $a + bi \neq 0$. If $a + bi = 0$, then $r = 0$ in 6.13, and θ may be a completely arbitrary angle.

We now introduce some appropriate terms in the following definition.

6.14 Definition. The expression on the right side of 6.13 is called the *trigonometric form* of the complex number $a + bi$. The nonnegative real number $r = \sqrt{a^2 + b^2}$ is called the *absolute value* of the complex number $a + bi$ and may be designated by $|a + bi|$. The angle θ occurring in 6.13 is called *an angle* of $a + bi$.

Clearly, the nonnegative real number r occurring in the trigonometric form of $a + bi$ is uniquely determined. However, the angle θ is not unique, but if $r \neq 0$ and θ_1 and θ_2 are any two possible angles of $a + bi$, then elementary properties of the sine and cosine functions show that $\theta_1 = \theta_2 + n \cdot 360°$ for some integer n.

As a consequence of these observations, let us point out that if r and s are positive real numbers and we know that

$$r(\cos \theta + i \sin \theta) = s(\cos \phi + i \sin \phi),$$

then necessarily $r = s$ and $\theta = \phi + n \cdot 360°$ for some integer n.

We have previously defined (3.7) absolute values for an ordered integral domain, and we know that the field of complex numbers is not ordered. However, the present definition of absolute value is an extension of the concept for real numbers. For if a is a real number, we may consider it to be the complex number $a + 0i$ and, by 6.14, we have $|a| = \sqrt{a^2}$. But if c is a positive real number, by $\sqrt{c}$ we mean the positive square root of c. It follows that $\sqrt{a^2} = a$ if $a \geq 0$, whereas $\sqrt{a^2} = -a$ if $a < 0$. Hence, for a *real* number a, the present meaning of $|a|$ coincides with its meaning according to Definition 3.7.

Let us now illustrate the trigonometric form of a complex number by some examples. First, let us consider the number $-2 + 2i$. As indicated in Figure 11, $|-2 + 2i| = 2\sqrt{2}$, and an angle of $-2 + 2i$ is 135°. Hence, 6.13 takes the form

$$-2 + 2i = 2\sqrt{2}(\cos 135° + i \sin 135°),$$

which is easily verified by direct calculation. Other examples, which the reader may check, are the following:

$$\begin{aligned} 1 + \sqrt{3}i &= 2(\cos 60° + i \sin 60°), \\ 4 &= 4(\cos 0° + i \sin 0°), \\ -i &= 1(\cos 270° + i \sin 270°), \\ -2(\cos 40° + i \sin 40°) &= 2(\cos 220° + i \sin 220°). \end{aligned}$$

It is only in special cases that we can find in degrees an angle of a given complex number. Naturally, an approximation may be obtained by use of trigonometric tables, or an angle may be merely indicated as in the following example. Let us attempt to express $1 + 3i$ in trigonometric form. Clearly, $|1 + 3i| = \sqrt{10}$, but we cannot exactly express its angle in degrees. However, if θ_1 is the positive acute angle such that $\tan \theta_1 = 3$, as indicated in Figure 12, we may write

$$1 + 3i = \sqrt{10}(\cos \theta_1 + i \sin \theta_1)$$

as the trigonometric form of $1 + 3i$.

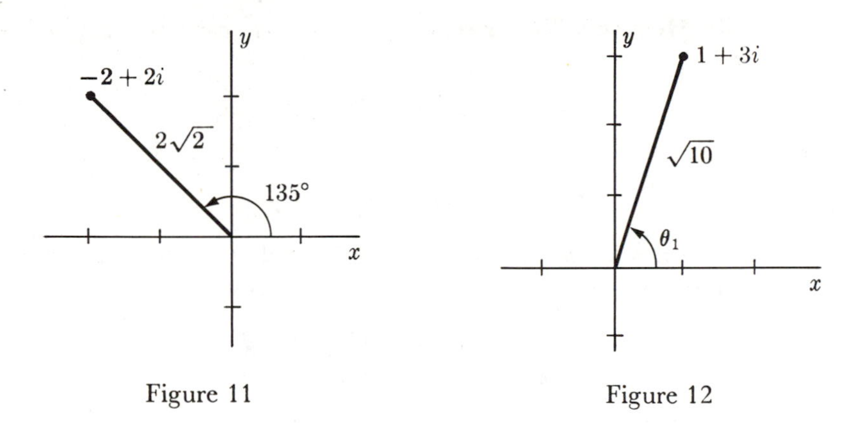

Figure 11

Figure 12

The fact that every complex number can be expressed in trigonometric form is of considerable significance largely because of the following remarkable theorem.

6.15 Theorem. *If u and v are complex numbers such that in trigonometric form*

$$u = r(\cos\theta + i\sin\theta)$$

and

$$v = s(\cos\phi + i\sin\phi),$$

then the trigonometric form of uv is given by

6.16 $$uv = rs(\cos(\theta+\phi) + i\sin(\theta+\phi)).$$

Otherwise expressed, $|uv| = |u|\cdot|v|$, and an angle of uv is the sum of an angle of u and an angle of v.

PROOF. To establish this result we need only multiply together the trigonometric forms of u and v and then use the simple addition formulas of trigonometry. Thus we have

$$\begin{aligned} uv &= rs(\cos\theta + i\sin\theta)(\cos\phi + i\sin\phi) \\ &= rs[(\cos\theta\cos\phi - \sin\theta\sin\phi) + i(\cos\theta\sin\phi + \sin\theta\cos\phi)] \\ &= rs[\cos(\theta+\phi) + i\sin(\theta+\phi)], \end{aligned}$$

and the desired result follows immediately.

The special case of the preceding theorem in which $u = v$ shows at once that

$$u^2 = r^2(\cos 2\theta + i\sin 2\theta).$$

The following generalization of this result is of great importance.

6.17 De Moivre's Theorem. *If n is an arbitrary positive integer and*

$$u = r(\cos \theta + i \sin \theta),$$

then

6.18
$$u^n = r^n(\cos n\theta + i \sin n\theta).$$

PROOF. For $n = 3$, we use Theorem 6.15 and the case in which $n = 2$ which has just been established as follows:

$$\begin{aligned} u^3 = u^2 \cdot u &= [r^2(\cos 2\theta + i \sin 2\theta)][r(\cos \theta + i \sin \theta)] \\ &= r^3(\cos 3\theta + i \sin 3\theta). \end{aligned}$$

A general proof along these lines can be easily given by induction.

One application of this theorem will be given in the next section. However, let us point out here how certain trigonometric identities can be obtained in an easy way by use of this theorem. By letting $r = 1$ and, as an example, taking $n = 3$, we see that

$$(\cos \theta + i \sin \theta)^3 = \cos 3\theta + i \sin 3\theta.$$

However, by actually multiplying out the left side, we find that

$$(\cos \theta + i \sin \theta)^3 = \cos^3 \theta - 3 \cos \theta \sin^2 \theta + i(3 \cos^2 \theta \sin \theta - \sin^3 \theta),$$

and it follows that

$$\cos 3\theta + i \sin 3\theta = \cos^3 \theta - 3 \cos \theta \sin^2 \theta + i(3 \cos^2 \theta \sin \theta - \sin^3 \theta).$$

From this equation we get at once the two following trigonometric identities:

$$\cos 3\theta = \cos^3 \theta - 3 \cos \theta \sin^2 \theta$$

and

$$\sin 3\theta = 3 \cos^2 \theta \sin \theta - \sin^3 \theta.$$

EXERCISES

1. Express each of the following complex numbers in trigonometric form and indicate the points in a coordinate plane that have these numbers as coordinates:

   (a) $-1 - i$, (b) $-\sqrt{3} + i$,
   (c) $\sqrt{3} + i$, (d) $-1 + \sqrt{3}i$,
   (e) -4, (f) $3 - 2i$,
   (g) $2 - 2i$, (h) $\cos 26° - i \sin 26°$.

2. Express each of the following complex numbers in the form $a + bi$:

(a) $4(\cos 45° + i \sin 45°)$,
(b) $2(\cos 120° + i \sin 120°)$,
(c) $3(\cos 180° + i \sin 180°)$,
(d) $3(\cos 270° + i \sin 270°)$,
(e) $\frac{1}{2}(\cos 300° + i \sin 300°)$,
(f) $12(\cos 0° + i \sin 0°)$,
(g) $11(\cos 90° + i \sin 90°)$,
(h) $(\cos 117° + i \sin 117°)(\cos 123° + i \sin 123°)$.

3. Use De Moivre's Theorem to compute each of the following, and then express your answers in algebraic form by evaluating the necessary trigonometric functions:

(a) $(-1 - i)^5$,
(b) $(\sqrt{3} - i)^8$,
(c) $(-i)^{12}$,
(d) $\left(\frac{1}{\sqrt{2}} + \frac{i}{\sqrt{2}}\right)^{100}$,
(e) $\left(-\frac{1}{2} - \frac{\sqrt{3}i}{2}\right)^6$,
(f) $(-1 + i)^{10}$,
(g) $(1 - \sqrt{3}i)^{11}$,
(h) $(\cos 18° + i \sin 18°)^{10}$.

4. Verify that the points with coordinates

$$(\cos 60° + i \sin 60°)^n \qquad (n = 1, 2, 3, 4, 5, 6)$$

are the vertices of a regular hexagon inscribed in a circle of radius 1.

5. If u^* is the conjugate of the complex number u, verify each of the following:

(a) $|u^*| = |u|$,
(b) $uu^* = |u|^2$,
(c) $u^{-1} = \frac{u^*}{|u|^2}$, if $u \neq 0$.

6. Show that if $u \neq 0$, De Moivre's Theorem also holds for every *negative* integer n.

7. Let $u, v \in \mathbf{C}$, and let P and Q be the points in a coordinate plane having respective coordinates u and v. Let R be the point with coordinate $u + v$. If O is the origin, show that OR is a diagonal of the parallelogram having OP and OQ as adjacent sides.

8. Show that if $u, v \in \mathbf{C}$, then $|u + v| \leq |u| + |v|$.

9. Use De Moivre's Theorem to find trigonometric identities for $\cos 4\theta$ and $\sin 4\theta$.

10. If $u = a + bi$, we have defined $|u| = \sqrt{a^2 + b^2}$. Use this definition to prove directly that if $u, v \in \mathbf{C}$, then $|uv| = |u| \cdot |v|$.

11. If $u, v \in \mathbf{C}$ with $v \neq 0$, prove that $\left|\dfrac{u}{v}\right| = \dfrac{|u|}{|v|}$.

6.6 THE *n*th ROOTS OF A COMPLEX NUMBER

In this section we give an important application of the use of the trigonometric form of a complex number. First, we give the following familiar definition.

6.19 Definition. Let n be a positive integer greater than 1. If $u, v \in \mathbf{C}$ such that $v^n = u$, we say that v is an nth root of u.

We shall now prove the following theorem.

6.20 Theorem. *If n is a positive integer greater than* 1, *and*

$$u = r(\cos \theta + i \sin \theta)$$

is a nonzero complex number in trigonometric form, there exist exactly n nth roots of u, namely,

$$\textbf{6.21} \qquad r^{1/n}\left(\cos \frac{\theta + k \cdot 360^\circ}{n} + i \sin \frac{\theta + k \cdot 360^\circ}{n}\right)$$

$$(k = 0, 1, \cdots, n-1).$$

Here $r^{1/n}$ represents the principal nth root of the positive real number r, that is, the positive real nth root of r whose existence is asserted in Theorem 6.6.

PROOF. Suppose that $v = s(\cos \phi + i \sin \phi)$ is an nth root of u. Then $v^n = u$ and De Moivre's Theorem assures us that

$$s^n(\cos n\phi + i \sin n\phi) = r(\cos \theta + i \sin \theta).$$

It follows that the absolute values of the two members of this equation are equal, and an angle of one must be equal to an angle of the other. Hence $s^n = r$, so that $s = r^{1/n}$. Moreover, $n\phi = \theta + k \cdot 360^\circ$ for some integer k, and it follows that $\phi = (\theta + k \cdot 360^\circ)/n$. We have therefore shown that every nth root v of u must be of the form

$$\textbf{6.22} \qquad v = r^{1/n}\left(\cos \frac{\theta + k \cdot 360^\circ}{n} + i \sin \frac{\theta + k \cdot 360^\circ}{n}\right)$$

for some integer k. Conversely, it is readily verified by De Moivre's Theorem that if v is given by 6.22, then $v^n = u$ for *every* choice of the integer k. The number of distinct nth roots of u is therefore the number of ways in which the integer k can be chosen in 6.22 so as to give distinct values of v. The angles obtained by letting k take the values $0, 1, \cdots, n - 1$ have distinct terminal sides, and this fact makes it almost obvious that these n values of k yield distinct values of v. Moreover, if t is an arbitrary integer, the Division Algorithm asserts that there exist integers q and r with $0 \leq r < n$ such that $t = qn + r$, and therefore

$$\frac{\theta + t\cdot 360^\circ}{n} = \frac{\theta + r\cdot 360^\circ}{n} + q\cdot 360^\circ.$$

It is then clear that the angle $(\theta + t\cdot 360^\circ)/n$ has the same terminal side as the angle $(\theta + r\cdot 360^\circ)/n$. Since $0 \leq r < n$, we see that all possible different values of v are obtained if in 6.22 we let k take the values $0, 1, \cdots, n - 1$. This completes the proof of the theorem.

As an example of the use of this theorem, let us find the fifth roots of the complex number $-2 + 2i$. First, we express this number in trigonometric form as follows:

$$-2 + 2i = 2^{3/2}(\cos 135^\circ + i \sin 135^\circ).$$

In the notation of the theorem, we have $r = 2^{3/2}$, $\theta = 135^\circ$, and $n = 5$. Accordingly, the fifth roots of $-2 + 2i$ are the following:

$$\begin{aligned} &2^{3/10}(\cos 27^\circ + i \sin 27^\circ), \\ &2^{3/10}(\cos 99^\circ + i \sin 99^\circ), \\ &2^{3/10}(\cos 171^\circ + i \sin 171^\circ), \\ &2^{3/10}(\cos 243^\circ + i \sin 243^\circ), \\ &2^{3/10}(\cos 315^\circ + i \sin 315^\circ). \end{aligned}$$

An interesting special case of Theorem 6.20 arises if we choose $u = 1$; hence $r = 1$ and $\theta = 0^\circ$. We state this case as follows.

6.23 Corollary. *The distinct nth roots of* 1 *are the complex numbers*

6.24 $$\cos \frac{k\cdot 360^\circ}{n} + i \sin \frac{k\cdot 360^\circ}{n} \qquad (k = 0, 1, \cdots, n - 1).$$

By De Moivre's Theorem, we have

$$\left(\cos \frac{360^\circ}{n} + i \sin \frac{360^\circ}{n}\right)^k = \cos \frac{k\cdot 360^\circ}{n} + i \sin \frac{k\cdot 360^\circ}{n}.$$

Hence the n distinct nth roots of 1, as given in 6.24, may all be expressed as powers of a certain nth root of 1. We have then the following alternate form of the preceding corollary.

6.25 Corollary. *Let us set*

6.26
$$w = \cos \frac{360°}{n} + i \sin \frac{360°}{n},$$

so that w is the nth root of 1 *having the smallest positive angle. Then the nth roots of* 1 *are the numbers*

6.27
$$w, w^2, w^3, \cdots, w^n = 1.$$

Since all nth roots of 1 have absolute value 1, they are coordinates of points on the circle with radius 1 and center the origin. Moreover, it is clear from 6.24 that they are the vertices of a regular polygon of n sides inscribed in this circle, with one vertex at the real number 1. This fact is of considerable importance in the study of the constructability of regular polygons with ruler and compass.

EXERCISES

1. Find the cube roots of 1 and express the answers in algebraic form. Draw a figure showing that these numbers are the coordinates of the vertices of a regular polygon of three sides (equilateral triangle).
2. Do the corresponding thing for the fourth roots of 1.
3. Do the corresponding thing for the eighth roots of 1.
4. Show that the sixth roots of 1 are the cube roots of 1 and their negatives.
5. Find the required roots and express the answers in algebraic form:
   (a) The cube roots of $-2 + 2i$.
   (b) The cube roots of $-8i$.
   (c) The fourth roots of -4.
   (d) The sixth roots of $-i$.
   (e) The fourth roots of $-1 - \sqrt{3}i$.
   (f) The square roots of $-1 + \sqrt{3}i$.
6. In each of the following, express the required roots in trigonometric form:
   (a) The fifth roots of 1.
   (b) The fourth roots of $-1 + i$.
   (c) The fourth roots of $\frac{1}{2} + \frac{\sqrt{3}i}{2}$.

(d) The sixth roots of $1 - i$.
(e) The square roots of $1 + 2i$.
(f) The fourth roots of $16(\cos 12° + i \sin 12°)$.

7. Show that if v is any one of the nth roots of the nonzero complex number u, and w is given by 6.26, then $v, wv, w^2v, \cdots, w^{n-1}v$ are all the nth roots of u.

8. Show that the multiplicative inverse of an nth root of 1 is also an nth root of 1.

9. If $t \in \mathbf{C}$ such that $t^n = 1$ but $t^m \neq 1$ for $0 < m < n$, t is called a *primitive* nth root of 1. Show each of the following:
(i) The number w, defined in 6.26, is a primitive nth root of 1.
(ii) If t is a primitive nth root of 1, then $1, t, t^2, \cdots, t^{n-1}$ are distinct and are all of the nth roots of 1.
(iii) If t is a primitive nth root of 1, then t^l is also a primitive nth root of 1 if and only if l and n are relatively prime.

NOTES AND REFERENCES

Proofs of the fundamental theorem (6.3) on the field of real numbers will be found in many texts on analysis as well as on algebra. Dedekind's approach is presented in full detail by Landau in his book referred to at the end of Chapter 3. For details of Cantor's method, see, for example, Dubisch [4].

Chapter 7

Groups

In all the algebraic systems studied so far we have always had two operations, namely, addition and multiplication. Later on we shall continue the study of these systems and, in particular, shall give many additional examples of rings and fields. However, we now proceed to study an important class of systems in which there is only one operation. As soon as the definition of a *group* is given in the next section it will be apparent that we already have many examples of groups from previous chapters, although we have not used this terminology. The theory of groups is an important part of modern algebra, and many books have been written on the subject. In this chapter we shall present only a few of the most fundamental properties of groups and give a number of examples that may serve to suggest the wide range of applications of the theory.

7.1 DEFINITION AND SIMPLE PROPERTIES

Let "$\circ$" be a binary operation defined on a nonempty set G. We recall that this statement only means that if (a, b) is any ordered pair of elements of G, then $a \circ b$ is a uniquely determined element of G. As a matter of fact, it is customary to call this operation either "addition" or "multiplication," and to use the familiar notation that is associated with these words. However, we shall first give the definition of a group, using the symbol "$\circ$" for the operation.

7.1 Definition. A nonempty set G on which there is defined a binary operation* "$\circ$" is called a *group* (with respect to this operation) provided the following properties are satisfied:

(i) If $a, b, c \in G$, then $(a \circ b) \circ c = a \circ (b \circ c)$ (*associative law*).
(ii) There exists an element e of G such that $e \circ a = a \circ e = a$ for every element a of G (*existence of an identity*).
(iii) If $a \in G$, there exists an element x of G such that $a \circ x = x \circ a = e$ (*existence of inverses*).

In Section 1.4 we called an element e, whose existence is asserted in (ii), an identity for the operation "$\circ$." However, we shall now call it an identity of the group. As suggested by the indicated name of the third property, as well as by previous use of the term, the element x whose existence is asserted in (iii) is called an *inverse* of the element a. Note that in a group *every* element has an inverse. As a matter of fact, it is quite easy to prove that in a group the identity is unique and also that every element has a unique inverse.

In order to give an example of a group, it is necessary to specify the elements of the set and to define an operation on this set in such a way that the three properties stated above are satisfied. We now give several examples of groups.

Example 1: The set **Z** of all integers, with the operation "$\circ$" taken as the usual operation $(+)$ of addition. The first property then merely states that

$$(a + b) + c = a + (b + c),$$

and this is just the associative law of addition for the integers. In this case, the identity of the group is the zero integer since $0 + a = a + 0 = a$ for $a \in Z$. The inverse of the element a is the element $-a$ since $a + (-a) = -a + a = 0$, and in the present notation this is just what is required in the statement of the third property. We have therefore verified all three properties, and hence we have a group. This group may be called the *additive group of the integers*.

This example can easily be generalized as follows. Let S be the set of all elements of any *ring*, and let the operation "$\circ$" be taken as the

* The assumption that "$\circ$" is a binary operation defined on G is sometimes expressed by saying that G is closed under this operation.

operation of addition already defined in the ring. Then the three properties of a group are precisely the properties P_2, P_3, and P_4 of Section 2.2 required of addition in the ring. Hence, S must be a group relative to the operation of addition. We shall refer to this group as the *additive group of the ring* S.

Example 2: The set T of all nonzero rational numbers, with the operation "$\circ$" taken as the familiar operation of multiplication of rational numbers. Since the product of two nonzero rational numbers is also a nonzero rational number, the set T is closed under multiplication; that is, multiplication is an operation defined on T. If a, b, $c \in T$, then $(ab)c = a(bc)$ by the associative law of multiplication for rational numbers, and this is just 7.1(i) in this case. Moreover, the identity is the rational number 1, and the inverse of an element a of T is the rational number a^{-1}. Hence, T is a group with respect to the operation of multiplication. We may emphasize that in a group *every* element must have an inverse, and this explains why the set of *all* rational numbers would not be a group with respect to multiplication.

This example can also be generalized as follows. If F is an arbitrary *field*, the set of all nonzero elements of F is a group with respect to the operation of multiplication in the field. This group we shall call the *multiplicative group of the field* F. We emphasize again that there is exactly one element of F, the zero, which is not an element of the multiplicative group of F.

Example 3: The set $\mathbf{R}^+$ of all *positive* real numbers with multiplication as the operation. Properties (ii) and (iii) are satisfied because $1 \in \mathbf{R}^+$ and if $a \in \mathbf{R}^+$, then also $a^{-1} \in \mathbf{R}^+$.

In like manner, $\mathbf{Q}^+$ is a group with respect to the operation of multiplication. More generally, if F is any *ordered* field, then F^+ is a group with respect to the operation of multiplication.

Example 4: The set $\{1, -1, i, -i\}$ consisting of these four complex numbers, with the operation of multiplication of complex numbers. It is easy to verify that the set is closed under multiplication and, of course, the number 1 is the identity. Moreover, 1 and -1 are their own inverses, and i and $-i$ are inverses of each other. The associative law clearly holds since it holds for multiplication of complex numbers in general.

Example 5: The set L of all complex numbers z with $|z| = 1$, again with respect to the operation of multiplication. Since, by Theorem 6.15, we have $|uv| = |u| \cdot |v|$, it follows that if $u \in L$ and $v \in L$, then $uv \in L$, and L is therefore closed under multiplication. Moreover, if $|u| = 1$, it is easy to verify that $|u^{-1}| = 1$. Hence, if $u \in L$, then also $u^{-1} \in L$ and each element of L has an inverse in L. The other properties are obviously satisfied, and therefore L is a group.

Example 6: Let H be the set $\{p, q, r\}$ with an operation, which we shall consider as multiplication, defined by the following table.

$(\cdot)$	p	q	r
p	p	q	r
q	q	r	p
r	r	p	q

Clearly, p is the identity of H. Moreover, p is its own inverse, and q and r are inverses of each other. The associative law is also satisfied, although it would be tedious to verify it from the table.

Example 7: Let G be the set $\{e, a, b, c\}$ with an operation of multiplication defined by the following table.

$(\cdot)$	e	a	b	c
e	e	a	b	c
a	a	e	c	b
b	b	c	e	a
c	c	b	a	e

The associative law holds, although we shall not verify it; and e is the identity. In this group, each element is its own inverse.

All of these examples have an additional property not required by the definition of a group. That is, they are abelian groups according to the following definition.

7.2 Definition. If in a group G with operation "$\circ$," $a \circ b = b \circ a$ for all $a, b \in G$, G is said to be an *abelian group* (or a commutative group).

The term "abelian group" is most commonly used for this concept. The name is derived from Niels Henrik Abel (1802–1829), a famous Norwegian mathematician whose fundamental work furnished an inspiration for many later mathematicians.

All the above examples are examples of abelian groups. In the following section we shall introduce some very important nonabelian groups.

As in the examples, we shall always call the operation in a group either addition or multiplication, and shall use the usual notation associated with these names. We shall never use addition as the operation in a nonabelian group. That is, whenever addition is used as the operation, we shall always assume, whether or not it is explicitly mentioned, that the group is abelian. It follows that in such a group *all* the properties of addition in a ring are satisfied. The identity will be denoted by 0 and called "zero"; the inverse of an element a will be denoted by $-a$; we shall write $b - a$ for $b + (-a)$, and so on.

When the operation in a group is called multiplication, the group may be either abelian or nonabelian. Accordingly, when we come to prove a property of arbitrary groups, we shall think of the operation as multiplication, and use the implied notation. In particular, the inverse of an element a will then be denoted by a^{-1}. We shall usually let e be the identity of the group, and reserve the symbol "1" for the smallest positive integer.

Now let G be an arbitrary group with operation multiplication. The following properties can be easily proved using only trivial modifications of proofs that we have already met in our study of rings and fields. We may emphasize that here multiplication need not be commutative. The first two of these properties have already been stated in the preceding section.

7.3 Theorem. *The following hold in every group G:*

(*i*) *The identity of G is unique.*
(*ii*) *If $a \in G$, a has a unique inverse a^{-1}.*
(*iii*) *If $a, b, c \in G$ such that $ab = ac$, then $b = c$.*
(*iv*) *If $a, b, c \in G$ such that $ba = ca$, then $b = c$.*
(*v*) *If $a, b \in G$, there exists a unique element x of G such that $ax = b$, and a unique element y of G such that $ya = b$. In fact, $x = a^{-1}b$ and $y = ba^{-1}$.*
(*vi*) *The inverse of a product is the product of the inverses in the reverse order, that is, if $a, b \in G$, then $(ab)^{-1} = b^{-1}a^{-1}$.*

Properties (iii) and (iv) are naturally called the *cancellation laws*. The proofs of the various parts of 7.3 will be assigned as an exercise below.

Just as in the case of multiplication in a ring, the generalized associative law holds, and we can write products without use of parentheses to indicate association.

If $a \in G$, we define $a^0 = e$, where e is the identity of the group. Then, just as though a were a nonzero element of a field, we can define a^n for *every* integer n. Moreover, for all choices of integers m and n, the following laws of exponents hold:

$$a^m \cdot a^n = a^{m+n},$$
$$(a^m)^n = a^{mn}.$$

For an *abelian* group, we also have $(ab)^n = a^n \cdot b^n$, but this is not true in general.

In those cases in which we use addition as the operation, we make use of multiples in place of powers; that is, na takes the place of a^n. Such a group is always assumed to be abelian and we have the following analogues of the above laws of exponents:

$$ma + na = (m + n)a,$$
$$n(ma) = (nm)a,$$
$$n(a + b) = na + nb.$$

These are properties that are already familiar as properties of addition in any ring.

A set H of elements of a group G is naturally called a *subgroup* of G if H is itself a group with respect to the operation already defined on G. If e is the identity of G, then G certainly has the two so-called *trivial* subgroups $\{e\}$ and G. Any other subgroup is called a *proper* subgroup.

The following theorem, whose proof will be required in Exercise 3 below, is often useful in determining subgroups of a given group.

7.4 Theorem. (*a*) *A nonempty subset K of a group G is a subgroup of G if and only if the following two conditions are satisfied:*

(*i*) *If $a, b \in K$, then $ab \in K$.*

(*ii*) *If $a \in K$, then $a^{-1} \in K$.*

(*b*) *If K has a finite number of elements, condition* (*ii*) *is implied by condition* (*i*).

In view of this theorem, we see therefore that a nonempty set of elements of a group G having a finite number of elements is a subgroup of G if and only if the set is closed under the operation on G. However, for infinite groups condition (ii) is not a consequence of condition (i). As an example, the set of all positive integers is a nonempty subset of the additive group of the integers which satisfies condition (i) but not condition (ii).

We conclude this section by introducing one additional concept which is analogous to the concept of direct sum of rings. Suppose that G and H are groups and let us first assume that they are abelian groups and that the operation in each of these groups is written as addition. On the Cartesian product set $G \times H$ we define an operation of addition as follows:

$$(g_1, h_1) + (g_2, h_2) = (g_1 + g_2, h_1 + h_2), \qquad g_1, g_2 \in G; h_1, h_2 \in H.$$

It is easy to verify that we obtain in this way an abelian group. This group is called the *direct sum* of G and H and usually denoted by $G \oplus H$.

In a similar manner, if G and H are completely arbitrary groups with the operation in both groups written as multiplication, we may modify the above procedure in the obvious way by defining mutliplication on the product set $G \times H$ as follows:

$$(g_1, h_1)(g_2, h_2) = (g_1g_2, h_1h_2), \qquad g_1, g_2 \in G; h_1, h_2 \in H.$$

The group obtained in this way is called the *direct product* of the groups G and H, and it is customary to denote it by the same notation $G \times H$ as used for the Cartesian product of the *sets* G and H.

We have defined the direct sum (or the direct product) of *two* groups. We leave it to the reader to give a corresponding definition of the direct sum (or the direct product) of any finite number of groups.

EXERCISES

1. Prove Theorem 7.3(i)–(vi).

2. Which of the following are groups with respect to the indicated operation?

(a) The set $\{1, 3, 7, 9\}$ of elements of $\mathbf{Z}_{10}$, with operation multiplication.

(b) The set $\{0, 2, 4, 6, 8\}$ of elements of $\mathbf{Z}_{10}$, with operation addition.

(c) The set $\{1, 3, 9\}$ of elements of $\mathbf{Z}_{10}$, with operation multiplication.

(d) The set of all rational numbers x such that $0 < x \leq 1$, with operation multiplication.

(e) The set of all positive irrational real numbers with operation multiplication.

(f) The set of all integers with operation "$\circ$" defined as follows: $a \circ b = a + b + 1$.

(g) The set of all integers with operation "$\circ$" defined as follows: $a \circ b = a - b$.

(h) The set of all rational numbers, other than 1, with operation "$\circ$" defined as follows: $a \circ b = a + b - ab$.

(i) The set of complex numbers that are nth roots of unity, where n is a fixed positive integer, with operation multiplication.

3. Prove Theorem 7.4. [Hint: For part (b), adapt the proof of Theorem 5.6.]

4. If H_1 and H_2 are subgroups of a group G, prove that $H_1 \cap H_2$ is a subgroup of G. Generalize this result by proving that the intersection of any number of subgroups of G is a subgroup of G.

5. Find all subgroups of each of the following groups:
(a) The additive group of the ring $\mathbf{Z}_{12}$.
(b) The additive group of the ring $\mathbf{Z}_5$.
(c) The multiplicative group of the field $\mathbf{Z}_7$.
(d) The multiplicative group of the field $\mathbf{Z}_{11}$.

6. Show that the set of all elements of the ring $\mathbf{Z}_n$ of the form $[k]$, where k and n are relatively prime, is a group with respect to the operation of multiplication.

7. Prove that $(ab)^2 = a^2b^2$ for all choices of a and b as elements of a group G if and only if G is abelian.

8. Let a be a fixed element of a group G. Prove that the set $\{x \mid x \in G, ax = xa\}$ is a subgroup of G.

9. If R is a ring with unity, prove that the set of all elements of R which have multiplicative inverses in R is a group G with respect to the operation of multiplication as defined in the ring R. Verify that the result of Exercise 6 above is a special case of this result.

10. If $R = \mathbf{Z}_4 \oplus \mathbf{Z}$, verify that the group G obtained by the method of the preceding exercise has four elements, and write out a multiplication table for this group.

11. Let G be the set of all ordered pairs (a, b) of real numbers with $a \neq 0$, and on this set let us define an operation of multiplication as follows:

$$(a, b)(c, d) = (ac, bc + d).$$

Verify that G is a nonabelian group.

7.2 MAPPINGS AND PERMUTATION GROUPS

Suppose that A, B, and C are sets and that we have given mappings $\alpha: A \to B$ and $\beta: B \to C$. It is then easy to define in a natural way a mapping of A into C. If $a \in A$, we first take the image $a\alpha$ of a under the mapping α. Now $a\alpha \in B$, so $(a\alpha)\beta$ is a uniquely determined element of C. Thus $a \to (a\alpha)\beta$, $a \in A$, defines a mapping of A into C determined by the given mappings α and β. We denote this mapping by $\alpha\beta$ and call it the *product* of α by β. More formally, the definition of the mapping $\alpha\beta: A \to C$ is as follows:

7.5 $$a(\alpha\beta) = (a\alpha)\beta, \qquad a \in A.$$

We may point out that, according to the definition just given, $\alpha\beta$ means "first perform α, then perform β." It is here that it makes an essential difference in notation whether we denote the image of a under the mapping α by $a\alpha$ or by the function notation $\alpha(a)$. Had we adopted the latter notation, the mapping which we have denoted by $\alpha\beta$ would map an element a of A into the element $\beta(\alpha(a))$ of C, and it would be natural to denote it by $\beta\alpha$. Both notations are widely used and in reading other books the student must be prepared to find either one.

As a simple illustration of Definition 7.5, let $A = \{1, 2, 3, 4\}$, $B = \{x, y, z\}$, $C = \{r, s\}$, and let $\alpha: A \to B$ and $\beta: B \to C$ be defined, respectively, as follows:

$$1\alpha = y, \quad 2\alpha = x, \quad 3\alpha = x, \quad 4\alpha = z;$$
$$x\beta = s, \quad y\beta = r, \quad z\beta = s.$$

Then the mapping $\alpha\beta: A \to C$ is obtained by the following calculations:

$$\begin{aligned} 1(\alpha\beta) &= (1\alpha)\beta = y\beta = r, \\ 2(\alpha\beta) &= (2\alpha)\beta = x\beta = s, \\ 3(\alpha\beta) &= (3\alpha)\beta = x\beta = s, \\ 4(\alpha\beta) &= (4\alpha)\beta = z\beta = s. \end{aligned}$$

It should be clear that in defining the product of two mappings a certain condition on the sets involved is necessary. Thus, if α is a mapping of A into B, $\alpha\beta$ is defined only if β is a mapping of the set B into some set.

We now take one more step as follows. Let A, B, C, and D be sets, and suppose that we have mappings $\alpha: A \to B$, $\beta: B \to C$, and $\gamma: C \to D$. Then $\alpha\beta$ is a mapping of A into C, and $(\alpha\beta)\gamma$ is a mapping of A into D. In like manner, $\alpha(\beta\gamma)$ is seen to be a mapping of A into D. It is an important fact that these two mappings are equal, that is, that

7.6 $$(\alpha\beta)\gamma = \alpha(\beta\gamma).$$

By the definition of equality of mappings, we shall prove 7.6 by verifying that

7.7 $$a((\alpha\beta)\gamma) = a(\alpha(\beta\gamma))$$

for every element a of A.

First, we observe that by the definition of the product of the mappings $\alpha\beta$ and γ, we have

$$a((\alpha\beta)\gamma) = (a(\alpha\beta))\gamma.$$

Then, by the definition of the product $\alpha\beta$, it follows that

$$(a(\alpha\beta))\gamma = ((a\alpha)\beta)\gamma,$$

and so the left side of 7.7 is equal to $((a\alpha)\beta)\gamma$. In like manner, by applying the definition of the product of α by $\beta\gamma$, and then the definition of the product $\beta\gamma$, we obtain

$$a(\alpha(\beta\gamma)) = (a\alpha)(\beta\gamma) = ((a\alpha)\beta)\gamma.$$

Since both sides of 7.7 are equal to $((a\alpha)\beta)\gamma$, we have proved 7.7 and also 7.6. Of course, Equation 7.6 merely states that *multiplication of mappings is always associative.*

In connection with our study of groups we are interested in the special case of mappings of a set A *onto* the *same set* A. Moreover, it is the one-one mappings of A onto A that we wish to study. The following terminology is convenient.

7.8 Definition. A one-one mapping of a set A onto itself is called a *permutation* of the set A.

The next theorem will show why we have paused to study mappings in a discussion of groups.

7.9 Theorem. *The set S of all permutations of a set A is a group with respect to the operation of multiplication of mappings defined in 7.5.*

PROOF. First, let us give a formal proof of the essentially obvious fact that if α and β are permutations of A, then $\alpha\beta$ is also a permutation of A. Since α and β are mappings of A onto A, the definition

$$a(\alpha\beta) = (a\alpha)\beta, \qquad a \in A,$$

of $\alpha\beta$ shows that $\alpha\beta$ is certainly a mapping of A *onto* A. We can now show that $\alpha\beta$ is, in fact, a one-one mapping and is therefore a permutation of A. We only need to verify that if $a, b \in A$ such that $a(\alpha\beta) = b(\alpha\beta)$, then $a = b$. However, $a(\alpha\beta) = b(\alpha\beta)$ can be written as $(a\alpha)\beta = (b\alpha)\beta$ and since β is a one-one mapping, we conclude that $a\alpha = b\alpha$. Then, since α is a one-one mapping, we have $a = b$, as we wished to show. This calculation shows that the set S is closed with respect to the operation of multiplication of permutations.

We have already proved that multiplication of mappings is associative, and hence the first requirement of a group is satisfied.

The identity mapping on the set A, as defined in Section 1.2, that is, the mapping $\epsilon: A \to A$ defined by

$$a\epsilon = a, \qquad a \in A,$$

is clearly an element of S and it is trivial to verify that $\alpha\epsilon = \epsilon\alpha = \alpha$ for every $\alpha \in S$. This shows that ϵ is the identity of our group. There only remains to prove that each element of S has an inverse.

If $\alpha \in S$, since α is a one-one mapping of A onto A, we have defined in Section 1.2 a one-one mapping α^{-1} of A onto A by the equation

7.10 $$(a\alpha)\alpha^{-1} = a, \qquad a \in A.$$

Thus $\alpha^{-1} \in S$ and, by our definition of product of mappings, we have that $\alpha\alpha^{-1} = \epsilon$. In order to prove that α^{-1} is indeed the inverse of α we must show that also $\alpha^{-1}\alpha = \epsilon$. However, we can verify this fact as follows, where a is an arbitrary element of A:

$$(a\alpha)(\alpha^{-1}\alpha) = a(\alpha(\alpha^{-1}\alpha)) = a((\alpha\alpha^{-1})\alpha) = a\alpha.$$

Since every element b of A is of the form $a\alpha$ for some $a \in A$, we have shown that

$$b(\alpha^{-1}\alpha) = b$$

for every element b of A. Hence $\alpha^{-1}\alpha = \epsilon$ and α^{-1}, defined by 7.10, is indeed the inverse of α.

We have now established all the properties required by the definition of a group, and the theorem is therefore established. This group S is naturally called the *group of all permutations of the set* A.

So far, the set A has been a completely arbitrary set. However, we are now primarily interested in the case in which A is restricted to have a finite number of elements. Accordingly, we make the following definition.

7.11 Definition. Let n be a positive integer. The group of all permutations of a set with n elements is called the *symmetric group* on n symbols, and may be designated by S_n.

Let us now consider an example in which $A = \{1, 2, 3\}$, a set with three elements. Then the symmetric group S_3, consisting of all permutations of A, contains the six elements $\alpha_1, \alpha_2, \alpha_3, \alpha_4, \alpha_5, \alpha_6$, as follows:

7.12

$$\begin{array}{lll} 1\alpha_1 = 1, & 2\alpha_1 = 2, & 3\alpha_1 = 3, \\ 1\alpha_2 = 2, & 2\alpha_2 = 1, & 3\alpha_2 = 3, \\ 1\alpha_3 = 3, & 2\alpha_3 = 2, & 3\alpha_3 = 1, \\ 1\alpha_4 = 1, & 2\alpha_4 = 3, & 3\alpha_4 = 2, \\ 1\alpha_5 = 2, & 2\alpha_5 = 3, & 3\alpha_5 = 1, \\ 1\alpha_6 = 3, & 2\alpha_6 = 1, & 3\alpha_6 = 2. \end{array}$$

The product of two of these permutations may, of course, be computed by using the definition of product of mappings. For example, let us compute $\alpha_2\alpha_5$. We have

$$\begin{aligned} 1(\alpha_2\alpha_5) &= (1\alpha_2)\alpha_5 = 2\alpha_5 = 3, \\ 2(\alpha_2\alpha_5) &= (2\alpha_2)\alpha_5 = 1\alpha_5 = 2, \\ 3(\alpha_2\alpha_5) &= (3\alpha_2)\alpha_5 = 3\alpha_5 = 1. \end{aligned}$$

Hence $\alpha_2\alpha_5 = \alpha_3$, since under the mapping $\alpha_2\alpha_5$ each element of A has the same image as under the mapping α_3. In like manner we can compute all products and obtain the following multiplication table for the group S_3.

7.13

	α_1	α_2	α_3	α_4	α_5	α_6
α_1	α_1	α_2	α_3	α_4	α_5	α_6
α_2	α_2	α_1	α_5	α_6	α_3	α_4
α_3	α_3	α_6	α_1	α_5	α_4	α_2
α_4	α_4	α_5	α_6	α_1	α_2	α_3
α_5	α_5	α_4	α_2	α_3	α_6	α_1
α_6	α_6	α_3	α_4	α_2	α_1	α_5

It is clear that α_1 is the identity of this group, and from the table one can easily find the inverse of each element. The group S_3 is not an abelian group since, for example, $\alpha_2\alpha_5 = \alpha_3$, whereas $\alpha_5\alpha_2 = \alpha_4$.

Before leaving this example, let us mention still another way of exhibiting the individual permutations of this group. For example, let us consider the element α_2 of S_3, as defined in 7.12. It is sometimes convenient to write

7.14

$$\alpha_2 = \begin{pmatrix} 1 & 2 & 3 \\ 2 & 1 & 3 \end{pmatrix}$$

to express the fact that under the mapping α_2 the image of 1 is 2, the image of 2 is 1, and the image of 3 is 3. According to this notation, we merely write the elements of the set A (in any order) in the top row, and under each element of A we write its image under the mapping α_2. In like manner, we see that

$$\alpha_5 = \begin{pmatrix} 1 & 2 & 3 \\ 2 & 3 & 1 \end{pmatrix}.$$

Then to compute the product $\alpha_2\alpha_5$ we observe that under this product 1 maps into 2 and then 2 maps into 3; hence 1 maps into 3. In like manner, 2 maps into 2 and 3 into 1. Hence, we may write

$$\begin{pmatrix} 1 & 2 & 3 \\ 2 & 1 & 3 \end{pmatrix}\begin{pmatrix} 1 & 2 & 3 \\ 2 & 3 & 1 \end{pmatrix} = \begin{pmatrix} 1 & 2 & 3 \\ 3 & 2 & 1 \end{pmatrix} = \alpha_3,$$

and we have again verified that $\alpha_2\alpha_5 = \alpha_3$.

We can use a notation similar to 7.14 to denote a permutation of any finite set. In general, if $\{i_1, i_2, \cdots, i_n\}$ is an arrangement of the integers $1, 2, \cdots, n$, then by

$$\begin{pmatrix} 1 & 2 & 3 & \cdots & n \\ i_1 & i_2 & i_3 & \cdots & i_n \end{pmatrix}$$

we mean the permutation α of the set $A = \{1, 2, \cdots, n\}$ such that $1\alpha = i_1$, $2\alpha = i_2$, $\cdots$, $n\alpha = i_n$. We shall use this notation whenever it seems convenient to do so.

We found above that S_3 has six elements. Let us now determine the number of elements in the symmetric group S_n, that is, the number of permutations of a set $A = \{1, 2, \cdots, n\}$ with n elements. Clearly, the image of 1 may be any element of A, and hence there are n choices for the image of 1. After an image of 1 is selected, there are then $n - 1$ choices for the image of 2, and so on. It follows that there are $n(n - 1) \cdot (n - 2) \cdots 2 \cdot 1$ different permutations of A. This number is usually denoted by $n!$ and called "n factorial." We have therefore shown that S_n has $n!$ elements.

Any group whose elements are permutations is naturally called a *permutation group* or a *group of permutations*. Any subgroup of a symmetric group S_n is certainly a permutation group. For example, from the table 7.13 and Theorem 7.4 it follows that $\{\alpha_1, \alpha_5, \alpha_6\}$ is a subgroup of S_3, and this is therefore an example of a permutation group which is not a symmetric group since it is not the group of *all* permutations of any set.

We conclude this section with a brief indication of how one can construct some interesting permutation groups by use of properties of symmetry of certain geometric figures. As an example, let us consider a square and study all rigid motions of the square into itself. That is, if the square is thought of as being made of some rigid material, such as cardboard, we consider motions such that the figure will look the same after the motion as before. In this, as well as in all other examples we shall consider, the rigid motions will consist of rotations either in the plane or in space. Each rigid motion of the square can be used in an almost obvious way to define a permutation of the vertices of the square. Let us designate the vertices of the square by 1, 2, 3, and 4. Moreover, let E, F, G, and H be the midpoints of the sides, as indicated in Figure 13; and let O be the center of the square. A rotation, in the plane of the square, through an angle of 90° about point O would place the vertices in the position shown in Figure 14. We may interpret the result of this

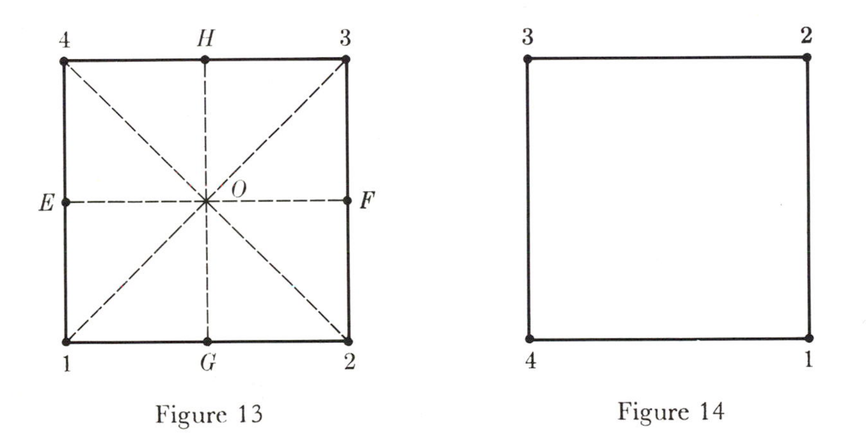

Figure 13 Figure 14

rotation as mapping 1 into 2, 2 into 3, 3 into 4, and 4 into 1, that is, as effecting the permutation

$$\alpha = \begin{pmatrix} 1 & 2 & 3 & 4 \\ 2 & 3 & 4 & 1 \end{pmatrix}$$

of the set $\{1, 2, 3, 4\}$ whose elements denote the vertices. A similar rotation through an angle of 180° or 270° leads to the respective permutations

$$\alpha^2 = \begin{pmatrix} 1 & 2 & 3 & 4 \\ 3 & 4 & 1 & 2 \end{pmatrix} \quad \text{or} \quad \alpha^3 = \begin{pmatrix} 1 & 2 & 3 & 4 \\ 4 & 1 & 2 & 3 \end{pmatrix}.$$

Clearly, $\alpha^4 = \epsilon$, the identity permutation. We also have other rigid motions consisting of rotations in space about a line of symmetry of the square. Let β be the permutation which arises from a rotation through an angle of 180° about the line EF and γ the permutation which arises from a similar rotation about GH. Then we see that

$$\beta = \begin{pmatrix} 1 & 2 & 3 & 4 \\ 4 & 3 & 2 & 1 \end{pmatrix} \quad \text{and} \quad \gamma = \begin{pmatrix} 1 & 2 & 3 & 4 \\ 2 & 1 & 4 & 3 \end{pmatrix}.$$

There remain two other permutations arising from rotations through 180° about the diagonals of the square. These are

$$\delta = \begin{pmatrix} 1 & 2 & 3 & 4 \\ 1 & 4 & 3 & 2 \end{pmatrix} \quad \text{and} \quad \sigma = \begin{pmatrix} 1 & 2 & 3 & 4 \\ 3 & 2 & 1 & 4 \end{pmatrix}.$$

The set $\{\epsilon, \alpha, \alpha^2, \alpha^3, \beta, \gamma, \delta, \sigma\}$ of permutations obtained in this way is closed under multiplication, as is easily verified by the multiplication table given below.

7.15

	ϵ	α	α^2	α^3	β	γ	δ	σ
ϵ	ϵ	α	α^2	α^3	β	γ	δ	σ
α	α	α^2	α^3	ϵ	σ	δ	β	γ
α^2	α^2	α^3	ϵ	α	γ	β	σ	δ
α^3	α^3	ϵ	α	α^2	δ	σ	γ	β
β	β	δ	γ	σ	ϵ	α^2	α	α^3
γ	γ	σ	β	δ	α^2	ϵ	α^3	α
δ	δ	γ	σ	β	α^3	α	ϵ	α^2
σ	σ	β	δ	γ	α	α^3	α^2	ϵ

Moreover, it is evident that each permutation of this set has an inverse in this set, and we therefore have a group of permutations. This particular group with eight elements is called the *octic group*. Since we obtained this group by a consideration of the rigid motions of a square, we may also say that it is the group of rigid motions of a square.

In a similar way we may construct the group of rigid motions of other geometric figures. We observe that by this process we must actually obtain a *group* of permutations. In the first place, a rigid motion followed by another rigid motion is itself a rigid motion, and hence the set we obtain must be closed under multiplication. Since, also, each rigid motion can be reversed by another rigid motion, the inverse of each permutation in the set will also be in the set. The fact that we obtain a group of permutations then follows from Theorem 7.4.

EXERCISES

1. In the following, α and β are the given permutations of the set $A = \{1, 2, 3, 4, 5\}$. Compute, in each case, $\alpha\beta$, $\beta\alpha$, α^2, and β^2.

(a) $1\alpha = 2,\quad 2\alpha = 1,\quad 3\alpha = 3,\quad 4\alpha = 5,\quad 5\alpha = 4;$
$1\beta = 1,\quad 2\beta = 4,\quad 3\beta = 2,\quad 4\beta = 3,\quad 5\beta = 5;$

(b) $1\alpha = 4,\quad 2\alpha = 3,\quad 3\alpha = 5,\quad 4\alpha = 1,\quad 5\alpha = 2;$
$1\beta = 2,\quad 2\beta = 3,\quad 3\beta = 1,\quad 4\beta = 4,\quad 5\beta = 5;$

(c) $1\alpha = 2,\quad 2\alpha = 1,\quad 3\alpha = 4,\quad 4\alpha = 5,\quad 5\alpha = 3;$
$1\beta = 2,\quad 2\beta = 3,\quad 3\beta = 4,\quad 4\beta = 5,\quad 5\beta = 1.$

(d) $\alpha = \begin{pmatrix} 1 & 2 & 3 & 4 & 5 \\ 5 & 4 & 3 & 1 & 2 \end{pmatrix}, \quad \beta = \begin{pmatrix} 1 & 2 & 3 & 4 & 5 \\ 3 & 2 & 1 & 5 & 4 \end{pmatrix}.$

(e) $\alpha = \begin{pmatrix} 1 & 2 & 3 & 4 & 5 \\ 1 & 3 & 2 & 5 & 4 \end{pmatrix}, \quad \beta = \begin{pmatrix} 1 & 2 & 3 & 4 & 5 \\ 2 & 3 & 1 & 4 & 5 \end{pmatrix}.$

(f) $\alpha = \begin{pmatrix} 1 & 2 & 3 & 4 & 5 \\ 5 & 4 & 3 & 2 & 1 \end{pmatrix}, \quad \beta = \begin{pmatrix} 1 & 2 & 3 & 4 & 5 \\ 5 & 4 & 2 & 1 & 3 \end{pmatrix}.$

2. Verify the entry in the table 7.15 giving each of the following products: $\alpha^2\sigma$, $\beta\gamma$, $\gamma\beta$, $\alpha^3\gamma$, $\delta\sigma$.

3. Find all subgroups of the symmetric group S_3.

4. Find all subgroups of the octic group (7.15).

5. Show that the group of rigid motions of an equilateral triangle is the symmetric group S_3.

6. Find the group of rigid motions of a rectangle that is not a square. Make a multiplication table for this group and show that it is a subgroup of the octic group.

7. How many elements are there in the group of rigid motions of a regular pentagon? A regular hexagon?

8. Let $\alpha: A \to B$ be a given mapping. Prove each of the following:

(i) If ϵ_A is the identity mapping on the set A, there exists a mapping $\beta: B \to A$ such that $\alpha\beta = \epsilon_A$ if and only if α is a one-one mapping.

(ii) There exists a mapping $\gamma: B \to A$ such that $\gamma\alpha = \epsilon_B$ if and only if α is an onto mapping.

(iii) If both the mappings β and γ exist, as defined in parts (i) and (ii), then $\beta = \gamma$.

7.3 HOMOMORPHISMS AND ISOMORPHISMS

The concept of a group homomorphism is essentially the same as that of a ring homomorphism except that we now have only one operation instead of two. However, the operations may be written differently in the two groups, so we shall state the definition in the following general way.

7.16 Definition. Let G be a group with operation "$\circ$" and H a group with operation "$\square$." A mapping $\theta: G \to H$ of G into H is called a *homomorphism* if and only if for $a, b \in G$, we have

7.17 $$(a \circ b)\theta = (a\theta) \square (b\theta).$$

If there exists a homomorphism of G *onto* H, we may say that G is *homomorphic* to H or that H is a *homomorphic image* of G.

Again, the special case of a homomorphism in which the mapping is one-one is of such importance that we introduce as follows the terminology which has already been used for rings.

7.18 Definition. A homomorphism which is a one-one mapping is called an *isomorphism.* If there exists an isomorphism of G *onto* H, we may say that G is *isomorphic* to H or that H is an *isomorphic image* of G.

Of course, just as with rings, if θ is an isomorphism of G onto H, then θ^{-1} is an isomorphism of H onto G, and we may say that G and H are isomorphic.

The operations "$\circ$" and "$\square$" appearing in Definition 7.16 will be considered to be either addition or multiplication, but the new feature here is that one of them may be written as addition and the other as multiplication. In any case, we may for convenience indicate that condition 7.17 holds by saying that *the group operation is preserved under the mapping* θ.

Let us now illustrate these concepts by a few illustrative examples.

Example 1: Let G be the additive group of the ring $\mathbf{Z}_4$ and H the multiplicative group of the field $\mathbf{Z}_5$. For convenience, let us designate the elements of G by 0, 1, 2, 3 and the elements of H by 1*, 2*, 3*, 4*. Now let $\theta: G \to H$ be the one-one mapping of G onto H defined as follows:

$$0\theta = 1^*, \quad 1\theta = 2^*, \quad 2\theta = 4^*, \quad 3\theta = 3^*.$$

We assert that the group operation is then preserved under the mapping θ. For example, we have

$$(2 + 3)\theta = 1\theta = 2^* = 4^* \cdot 3^* = (2\theta)(3\theta).$$

The reader may verify that in every case a sum of elements of G has as image the product of the corresponding images in H. Hence θ is an isomorphism of G onto H.

Example 2: Let $\mathbf{R}^+$ be the group of all positive real numbers with operation multiplication (Example 3 of Section 7.1) and let L be the additive group of the field of all real numbers. If $\phi: \mathbf{R}^+ \to L$ is defined by

$$x\phi = \log_{10} x, \qquad x \in \mathbf{R}^+,$$

it is known that ϕ is a one-one mapping of $\mathbf{R}^+$ onto L and, moreover, one of the familiar laws of logarithms assures us that

$$(xy)\phi = \log_{10} xy = \log_{10} x + \log_{10} y = x\phi + y\phi, \quad x, y \in \mathbf{R}^+.$$

This shows that the group operation is preserved under the mapping ϕ, and ϕ is therefore an isomorphism of $\mathbf{R}^+$ onto L. It is this fact that is of central importance in the theory of logarithms. Of course, in place of 10 we could use as a base any fixed element of $\mathbf{R}^+$ other than 1.

Example 3: Let S_3 be the symmetric group on three symbols, with multiplication table given by 7.13, and let H be the group consisting of the set $\{1, -1, i, -i\}$ of complex numbers with operation multiplication. Let $\theta: S_3 \to H$ be defined as follows:

$$\alpha_1\theta = 1, \quad \alpha_2\theta = -1, \quad \alpha_3\theta = -1, \quad \alpha_4\theta = -1, \quad \alpha_5\theta = 1, \quad \alpha_6\theta = 1.$$

It may be verified that the group operation is preserved under this mapping. As an example, we have

$$(\alpha_2\alpha_4)\theta = \alpha_6\theta = 1 = (-1)(-1) = (\alpha_2\theta)(\alpha_4\theta).$$

Thus θ is a homomorphism of S_3 into H. However, θ is clearly not an onto mapping, so we cannot say that H is a homomorphic image of S_3. However, if we let $H' = \{1, -1\}$, then H' is a subgroup of H and θ does define a mapping of S_3 onto H', and hence H' is a homomorphic image of S_3.

Example 4: Let G and H be arbitrary groups with operation multiplication. Then the mapping $\theta: G \times H \to G$ defined by $(g, h)\theta = g$ (that is, the projection of the set $G \times H$ onto G) is clearly a homomorphism of the direct product $G \times H$ onto G. Similarly, there exists a homomorphism of $G \times H$ onto H. Thus, both of the groups G and H are homomorphic images of the direct product $G \times H$.

We may point out that there always exists a (trivial) homomorphism of an arbitrary group G into an arbitrary group H in which every element of G maps into the identity of H.

Some fundamental properties of homomorphisms of groups are stated in the following theorem (cf. Theorem 2.33), in which we shall consider the operations of both groups to be multiplication.

7.19 Theorem. *Let $\theta: G \to H$ be a homomorphism of the group G into the group H. Then each of the following is true:*

(*i*) *If e is the identity of G, then $e\theta$ is the identity of H.*
(*ii*) *If $a \in G$, then $(a^{-1})\theta = (a\theta)^{-1}$.*
(*iii*) *If G is abelian and θ is an onto mapping, then H is abelian.*

We leave the proof of this theorem as an exercise.

The reader may sometimes have to make a suitable modification in notation. For example, if the operation in G is multiplication and that in H is addition, property (ii) of the preceding theorem should be interpreted as stating that $(a^{-1})\theta = -(a\theta)$, since $-(a\theta)$ is the inverse of $a\theta$ in H.

The concept introduced in the following definition plays an important role in the study of homomorphisms of groups.

7.20 Definition. Let $\theta: G \to H$ be a homomorphism of the group G into the group H. The set of all elements of G which map into the identity of H is called the *kernel* of the homomorphism θ, and may be denoted by ker θ.

In Example 3 on p. 159, ker $\theta = \{\alpha_1, \alpha_5, \alpha_6\}$. It may be verified that in this case ker θ is actually a subgroup of S_3. The next theorem will show that this fact is no accident.

7.21 Theorem. *If $\theta: G \to H$ is a homomorphism, then ker θ is a subgroup of G. Moreover, if e is the identity of G, ker $\theta = \{e\}$ if and only if θ is an isomorphism.*

PROOF. Let $K = \ker \theta$. By Theorem 7.19(i), the identity of H is $e\theta$. If $a, b \in K$, then $(ab)\theta = (a\theta)(b\theta) = (e\theta)(e\theta) = e\theta$, and thus K is closed with respect to multiplication. Moreover, if $a \in K$, by Theorem 7.19(ii) we have that $(a^{-1}\theta) = (a\theta)^{-1} = (e\theta)^{-1} = e\theta$ and hence $a^{-1} \in K$. Theorem 7.4 then shows that K is a subgroup of G.

To prove the second statement of the theorem, suppose first that θ is an isomorphism. Since $e \in \ker \theta$, and the mapping is one-one, this assures us that no other element of G has as image the identity $e\theta$ of H, and it follows that ker $\theta = \{e\}$. Conversely, let us assume that θ is a homomorphism of G into H such that ker $\theta = \{e\}$. Suppose that $a, b \in G$ such that $a\theta = b\theta$. It follows that

$$(ab^{-1})\theta = (a\theta)(b^{-1}\theta) = (a\theta)(b\theta)^{-1} = e\theta,$$

and hence that $ab^{-1} \in K$. Thus, $ab^{-1} = e$ and $a = b$. This shows that θ is a one-one mapping and is therefore an isomorphism. This completes the proof of the theorem.

The result just established is often useful in the following way. To show that a mapping of a group G into a group H is an isomorphism, we may first show that it is a homomorphism and then verify that its kernel consists of the identity only.

We conclude this section by proving a theorem which is due to the famous English mathematician, Arthur Cayley (1821–1895).

7.22 Theorem. *Every group G is isomorphic to a group of permutations.*

PROOF. In order to prove this result we need first of all to determine the *set* some of whose permutations we shall associate with the elements of the given group G. We make what is perhaps the most obvious choice, namely, the set of elements of G itself. Moreover, we shall let G denote both the group and the set of its elements as it suits our convenience. Actually, the desired permutations will be obtained by multiplication by the elements of the group. More precisely, let us first observe that if $a \in G$, then

$$\{xa \mid x \in G\} = G.$$

Therefore the mapping $\theta_a: G \to G$ defined by $x\theta_a = xa$, $x \in G$, is a mapping of G onto G and it is clearly also a one-one mapping; hence θ_a is a permutation of the set G associated with the element a of the group G.

Now let us set

$$H = \{\theta_a \mid a \in G\};$$

that is, H is the set of all permutations of the type introduced above. Since for $x \in G$, we have

$$x(\theta_a \cdot \theta_b) = (x\theta_a)\theta_b = (x\theta_a)b = xab,$$

we see that

7.23 $$\theta_a \cdot \theta_b = \theta_{ab},$$

and H is therefore closed with respect to multiplication. Moreover, H has identity θ_e, where e is the identity of G; and 7.23 shows that $\theta_{a^{-1}}$ is the inverse of θ_a. Hence H is a subgroup of the group of all permutations of the set G.

We now assert that the mapping $\alpha: G \to H$ defined by $a\alpha = \theta_a$, $a \in G$, is an isomorphism of G onto H. It is clearly an onto mapping. Moreover, it is a homomorphism since, by 7.23,

$$(ab)\alpha = \theta_{ab} = \theta_a \cdot \theta_b = (a\alpha)(b\alpha).$$

There remains only to prove that α is a one-one mapping. However, if $\theta_a = \theta_b$, it follows that $xa = xb$ for every element x of G, and clearly we must have $a = b$. Thus α is indeed a one-one mapping, and this completes the proof.

In view of this theorem, in order to prove a theorem for arbitrary groups it is sufficient to prove it for groups of permutations. Although the subject of group theory is sometimes approached through the study of permutation groups, we shall not limit ourselves to this point of view.

EXERCISES

1. Verify that there exists an isomorphism of the additive group of the ring $\mathbf{Z}_6$ onto the multiplicative group of the field $\mathbf{Z}_7$ such that the image of the element 1 of the first group is the element 3* of the second group.
2. What familiar properties of logarithms follow from Theorem 7.19 as applied to the isomorphism given in Example 2 on p. 159?
3. Prove that a group G is abelian if and only if the mapping $\theta: G \to G$ defined by $a\theta = a^{-1}$, $a \in G$, is an isomorphism.
4. If G is the additive group of the ring $\mathbf{Z}_{15}$ and H is the additive group of the ring $\mathbf{Z}_5$, find a homomorphism of G onto H.
5. Show that two groups are necessarily isomorphic if each of them has exactly two elements. Show that the same conclusion holds if each has exactly three elements.
6. Find two groups, each with exactly four elements, which are not isomorphic. Do the same thing, with "four" replaced by "six."
7. If G is the multiplicative group of the field $\mathbf{Z}_5$, use the method of proof of Theorem 7.22 to find a group of permutations which is isomorphic to G.
8. If $\theta: G \to H$ and $\phi: H \to K$ are group homomorphisms, show that $\theta\phi$ is also a homomorphism.

9. The subgroup $\{\epsilon, \alpha^2, \beta, \gamma\}$ of the octic group (7.15) is sometimes called the *four-group*. Verify that the four-group is isomorphic to the additive group of the ring $\mathbf{Z}_2 \oplus \mathbf{Z}_2$, and also to the group of Example 7 of Section 7.1.

10. If $\theta: G \to H$ is a homomorphism of G into H and $a \in G$, prove that $a^k\theta = (a\theta)^k$ for *every* integer k.

11. Verify that the group G of Exercise 11 at the end of Section 7.1 has a subgroup which is isomorphic to the multiplicative group of the field of real numbers.

12. For each ordered pair (a, b) of real numbers with $a \neq 0$, let $\alpha_{a,b}: \mathbf{R} \to \mathbf{R}$ be the mapping of the field $\mathbf{R}$ into the field $\mathbf{R}$ defined by $x\alpha_{a,b} = ax + b$, $x \in \mathbf{R}$. Prove that the set H of all such mappings $\alpha_{a,b}$ is a group of permutations of $\mathbf{R}$, and that H is isomorphic to the group G mentioned in the preceding exercise.

13. Determine the set L of all elements $\alpha_{a,b}$ of the group H of the preceding exercise such that $\alpha_{a,b}\alpha_{1,2} = \alpha_{1,2}\alpha_{a,b}$. How do you know without detailed calculation that L is a subgroup of H? Verify that the group L is isomorphic to the additive group of the field $\mathbf{R}$.

14. Let G be a group and H merely a *set* on which a binary operation is defined. If there exists a mapping $\theta: G \to H$ of G *onto* H which preserves the operation, prove that H is a group with respect to the given operation and that θ is a homomorphism of the group G onto the group H. [This fact is usually expressed by saying that a homomorphic image of a group is a group.]

15. Let G be the multiplicative group of the field $\mathbf{Q}$, and let H be the set of all rational numbers other than 1. If $a, b \in H$, let us define $a \circ b = a + b - ab$. Prove that H is a group with respect to this operation. Prove also that the mapping $\theta: G \to H$ defined by $a\theta = 1 - a$, $a \in G$, is an isomorphism of G onto H.

7.4 CYCLIC GROUPS

If a is an element of an arbitrary group G, then since G is closed with respect to the operation (which we will consider to be multiplication), we see that $a^k \in G$ for every positive integer k. Moreover, a^0 is the identity e of G by definition, and a^{-k} is the inverse of a^k. It follows easily that the set $\{a^k \mid k \in \mathbf{Z}\}$ is a subgroup of G. We are particularly interested in the case in which this subgroup turns out to be all of G. Accordingly, let us make the following definition.

7.24 Definition. If the group G contains an element a such that $G = \{a^k \mid k \in \mathbf{Z}\}$, we say that G is a *cyclic group* and that G is *generated by* a or that a is a *generator of* G.

Since $a^i \cdot a^j = a^j \cdot a^i$ for $i, j \in \mathbf{Z}$, we see that a cyclic group is necessarily abelian.

Whether or not a group G is cyclic, if $a \in G$, the subgroup $\{a^k \mid k \in \mathbf{Z}\}$ of G is a cyclic group which we naturally call the *subgroup of* G *generated by* a.

Let us now give some examples of cyclic groups.

Example 1: The multiplicative group of the field $\mathbf{Z}_5$. Let us write the elements as 1, 2, 3, and 4; and remember that multiplication is to be carried out modulo 5. It is easily verified that $2^1 = 2$, $2^2 = 4$, $2^3 = 3$, $2^4 = 1$; hence each element of the group is of the form 2^k for some integer k. It follows that the group must be cyclic with generator 2. The reader may show that 3 is also a generator of this group. The element 4 generates the cyclic subgroup $\{1, 4\}$.

Example 2: The additive group of the ring $\mathbf{Z}$ of integers. In a ring with addition as the operation, ka is the analogue of a^k used above. The integer 1 is a generator of this group since every element is of the form $k \cdot 1$ for some integer k.

Example 3: The additive group of the ring $\mathbf{Z}_n$ of integers modulo n. This group is generated by the element 1 of $\mathbf{Z}_n$. Of course, it may have other generators as well.

Example 4: The group of all complex numbers that are nth roots of unity, where n is a fixed positive integer, with multiplication as the operation. The fact that this group is cyclic is the content of Corollary 6.25, where it was shown that $w = \cos(360°/n) + i \sin(360°/n)$ is a generator.

Example 5: The subgroup $\{\alpha_1, \alpha_5, \alpha_6\}$ of the symmetric group S_3, whose multiplication table is given in 7.13. It is easily verified that $\alpha_5^2 = \alpha_6$ and $\alpha_5^3 = \alpha_1$; hence α_5 is a generator of this subgroup. As a matter of fact, α_6 is also a generator of this same subgroup. The group S_3 is not a cyclic group. This may be shown by direct calculation or by observing that S_3 is nonabelian, whereas a cyclic group must be abelian.

We are now ready to give another definition as follows:

7.25 Definition. (i) If a group G has n elements, where n is a positive integer, G is said to have *finite order* or, more precisely,

to have *order n*. If there exists no such positive integer, G is said to have *infinite order*.
(ii) The *order of an element a* of a group G is the order of the cyclic subgroup of G generated by a.

In the language here introduced, we may say that the additive group of the integers has infinite order, the symmetric group S_n has order $n!$, the additive group of the ring $\mathbf{Z}_n$ has order n, and the multiplicative group of the field $\mathbf{Z}_p$ has order $p - 1$. All of these, except the first, are groups of finite order.

The next theorem gives an important characterization of the order of an element of a group. In fact, in order to compute the order of a given element it is usually simpler to apply this theorem than to use the definition.

7.26 Theorem. *An element a of a group G has order n if and only if n is the smallest positive integer such that $a^n = e$, where e is the identity of G. If no such integer exists, a has infinite order.*

As a first step in the proof, we shall prove the following lemma.

7.27 Lemma. *Let a be an element of the group G, and suppose that $a^n = e$, with n the smallest such positive integer. If $k \in \mathbf{Z}$, then $a^k = e$ if and only if $k \equiv 0 \pmod{n}$. More generally, if $i, j \in \mathbf{Z}$, then $a^i = a^j$ if and only if $i \equiv j \pmod{n}$.*

PROOF. By the Division Algorithm, we may write any integer k in the form $k = qn + r$, where q and r are integers and $0 \leq r < n$. Then, since $a^n = e$, we have

$$a^k = a^{qn+r} = (a^n)^q \cdot a^r = e^q \cdot a^r = a^r.$$

If $a^k = e$, we see that $a^r = e$, and it follows that $r = 0$ since, otherwise, r would be a positive integer less than n and we have assumed that n is the smallest positive integer such that $a^n = e$. Hence, if $a^k = e$, we have $k = qn$ and $k \equiv 0 \pmod{n}$. Conversely, if $k = qn$, it is apparent that $a^k = (a^n)^q = e^q = e$. This establishes the first part of the lemma. The second part now follows easily. For if $a^i = a^j$, it follows that $a^{i-j} = e$, and by what we have just proved, this is true if and only if $i - j \equiv 0 \pmod{n}$ or $i \equiv j \pmod{n}$.

Let us return to the proof of the theorem, and suppose first that $a^n = e$, with n as the smallest such positive integer. We now assert that the elements

7.28 $$e, a, a^2, \cdots, a^{n-1}$$

are distinct and are all of the elements of the cyclic subgroup of G generated by a. Since no two of the integers $0, 1, 2, \cdots, n-1$ are congruent modulo n, the preceding lemma shows that the elements 7.28 are distinct. Moreover, since every integer is congruent modulo n to some one of the integers $0, 1, 2, \cdots, n-1$, it also follows that a^k is equal to one of the elements 7.28, for every integer k. Hence the cyclic subgroup of G generated by a has exactly the n distinct elements 7.28; that is, it has order n and therefore a has order n.

To prove the converse, let us now assume that a has order n. Then not all positive powers of a can be distinct; that is, we must have $a^i = a^j$ for different positive integers i and j. Suppose that $i > j$, and it then follows that $a^{i-j} = e$, with $i - j > 0$. Hence, there exists some positive power of a which is equal to e. Suppose that m is the smallest positive integer such that $a^m = e$. Now, by what we have proved above, a has order m. Since it was given that a has order n, we must have $m = n$. This completes the proof of the first sentence of the theorem.

If there exists no positive integer n such that $a^n = e$, it is easy to show that $a, a^2, a^3, \cdots$ must all be distinct. (Why?) Hence a must have infinite order, and the theorem is established.

Theorem 7.26 makes it easy to determine the order of an element of a given group. For example, let us find the order of the element 3 of the multiplicative group of the ring $\mathbf{Z}_{11}$. By computing the successive powers of 3, we find that $3^2 = 9$, $3^3 = 5$, $3^4 = 4$, $3^5 = 1$. Hence, the element 3 has order 5. As another example, let us find the order of the element 10 of the additive group of the ring $\mathbf{Z}_{18}$. By Theorem 7.26, with the proper change of notation, this order will be the least positive integer n such that $n \cdot 10 \equiv 0 \pmod{18}$. It follows easily that $n = 9$, and the element 10 therefore has order 9.

In view of Theorems 7.26 and 5.8, we see that if a ring R has a unity e and has nonzero characteristic, the characteristic of R is simply the order of e in the additive group of R.

It is not difficult to prove that two cyclic groups are isomorphic if and only if they have the same order. This fact will follow immediately from the following theorem.

7.29 Theorem. (*i*) *Every cyclic group of infinite order is isomorphic to the additive group of the ring* $\mathbf{Z}$ *of integers.*
(*ii*) *Every cyclic group of order n is isomorphic to the additive group of the ring* $\mathbf{Z}_n$ *of integers modulo n.*

PROOF. First, let G be an infinite cyclic group with generator a, and let $\theta: \mathbf{Z} \to G$ be the mapping of $\mathbf{Z}$ into G defined by

$$k\theta = a^k, \qquad k \in \mathbf{Z}.$$

Then, if $i, j \in \mathbf{Z}$, we have

$$(i + j)\theta = a^{i+j} = a^i \cdot a^j = (i\theta)(j\theta),$$

and hence θ is a homomorphism of the additive group of the integers into the group G. Suppose, now, that $i, j \in \mathbf{Z}$ such that $i\theta = j\theta$, that is, such that $a^i = a^j$. If, for example, $i > j$, then $a^{i-j} = e$, the identity of G. But Theorem 7.26 would then show that a has finite order, whereas we are given that it has infinite order. Hence $i\theta = j\theta$ implies that $i = j$, and θ is a one-one mapping. Finally, every element of G is of the form a^k for some $k \in \mathbf{Z}$ and the mapping θ is therefore an onto mapping. This completes the proof of the first part of the theorem.

Next, let G be a cyclic group of order n with generator a. In order to keep the notation straight, we shall now denote the elements of $\mathbf{Z}_n$ by $[k]$, $k \in \mathbf{Z}$, and we shall show that the mapping $\theta: \mathbf{Z}_n \to G$ defined by

$$[k]\theta = a^k, \qquad k \in \mathbf{Z},$$

is the desired isomorphism of the additive group of $\mathbf{Z}_n$ onto the group G. First, let us show that the mapping is well-defined. In other words, let us show that if $i, j \in \mathbf{Z}$, then $[i] = [j]$ if and only if $a^i = a^j$. However, we know that $[i] = [j]$ if and only if $i \equiv j \pmod{n}$ and, by Theorem 7.26 and Lemma 7.27, we also know that $a^i = a^j$ if and only if $i \equiv j \pmod{n}$. Thus θ is indeed a mapping of $\mathbf{Z}_n$ into G, and it is clear that it is an onto mapping. Moreover, it is a one-one mapping since we have just observed that $a^i = a^j$ if and only if $[i] = [j]$. Finally we have

$$([i] + [j])\theta = [i + j]\theta = a^{i+j} = a^i \cdot a^j = ([i]\theta)([j]\theta).$$

Hence the operation is preserved, and θ is the required isomorphism of the additive group of $\mathbf{Z}_n$ onto the group G. The proof of the theorem is therefore complete.

In Example 1 above, we verified that the multiplicative group of the ring $\mathbf{Z}_5$ is cyclic with generator 2 (or 3), and clearly its order is 4. According to the theorem just proved, it must therefore be isomorphic to the additive group of the ring $\mathbf{Z}_4$. In order to distinguish between them, let us designate the elements of $\mathbf{Z}_4$ by [0], [1], [2], [3] and the nonzero elements of $\mathbf{Z}_5$ by 1, 2, 3, 4. Using the generator 2 of the latter group, the proof just given shows that $[k]\theta = 2^k$, $k \in \mathbf{Z}$, defines an isomorphism of these two groups. Written out in detail, this mapping is as follows:

$$[1]\theta = 2, \quad [2]\theta = 4, \quad [3]\theta = 3, \quad [0]\theta = 1.$$

We can obtain a different isomorphism by using the generator 3 of the multiplicative group of $\mathbf{Z}_5$. This isomorphism ϕ defined by $[k]\phi = 3^k$ yields the following explicit isomorphism:

$$[1]\phi = 3, \quad [2]\phi = 4, \quad [3]\phi = 2, \quad [0]\phi = 1.$$

We have thus exhibited two different isomorphisms of these two groups.

Our final theorem about cyclic groups is the following.

7.30 Theorem. *Every subgroup of a cyclic group G is itself a cyclic group.*

PROOF. Suppose that G is generated by a, and let H be a subgroup of G. Let m be the smallest positive integer such that $a^m \in H$. We shall show that H is a cyclic group generated by a^m. Since $H \subseteq G$, any element of H is of the form a^k for some integer k. By the Division Algorithm, we may write $k = qm + r$, where $0 \leq r < m$. Hence,

$$a^k = a^{qm+r} = (a^m)^q \cdot a^r,$$

and from this it follows that

$$a^r = (a^m)^{-q} \cdot a^k.$$

Since $a^m \in H$ and $a^k \in H$, this equation implies also that $a^r \in H$. In view of the choice of m as the smallest positive integer such that $a^m \in H$, and since $r < m$, we must have $r = 0$. We conclude that $k = qm$, and hence that every element a^k of H is of the form $(a^m)^q$ for some integer q. This shows that H is a cyclic group generated by a^m.

As an almost immediate consequence of the *proof* of the preceding theorem, we obtain the following result.

7.31 Corollary. *If a cyclic group G has finite order n and is generated by a, every subgroup H of G is generated by an element of the form a^m, where m is a divisor of n.*

PROOF. Since Theorem 7.26 shows that $a^n = e$, and $e \in H$, we apply the above argument with $k = n$ and obtain $n = qm$. Hence, m is a divisor of n.

Of course, by a simple change in notation, these results apply equally well to the case in which the operation is addition. As an illustration of the preceding corollary, let us find all subgroups of the additive group of the ring $\mathbf{Z}_{14}$. This is a cyclic group of order 14 generated by the element 1; hence the only subgroups are the cyclic subgroups generated by 1, 2, 7, and 14. The subgroup generated by 14 consists only of the identity 0. The subgroup generated by 2 has order 7 and the subgroup generated by 7 has order 2.

EXERCISES

1. Find the order of each element of the octic group (7.15).
2. Find an element of the symmetric group S_4 of order 4. Similarly, find an element of S_5 of order 5; of S_n of order n.
3. It can be proved that for every prime p, the multiplicative group of the field $\mathbf{Z}_p$ is cyclic. Verify this fact for $p = 7$, 11, and 13.
4. Find all subgroups of the additive group of the ring $\mathbf{Z}_{20}$.
5. Prove: If G is a cyclic group of order n and k is a positive divisor of n, there exists a subgroup of G of order k.
6. If G is a cyclic group of order n with generator a, prove that a^k is also a generator of G if and only if $(k, n) = 1$.
7. Apply the result of the preceding exercise to find all generators of the additive group of the ring $\mathbf{Z}_{30}$.
8. Prove that if $\theta: G \to H$ is a homomorphism of a cyclic group G with generator a onto a group H, then H is a cyclic group with generator $a\theta$ and that if G has finite order, the order of $a\theta$ is a divisor of the order of a.
9. Let G and H be cyclic groups of the same order, and let g be an arbitrary generator of G and h an arbitrary generator of H. Show that there exists an isomorphism θ of G onto H such that $g\theta = h$.
10. Determine all isomorphisms of the multiplicative group of the field $\mathbf{Z}_{11}$ onto the additive group of the ring $\mathbf{Z}_{10}$.
11. Determine all subgroups of the additive group of the ring $\mathbf{Z}$.

12. If a and b are elements of a group G such that a, b, and ab all have order 2, prove each of the following:
(i) $ab = ba$.
(ii) The set $\{e, a, b, ab\}$ is a subgroup of G.

13. Prove that if a and b are elements of a group G, then ab and ba have the same order.

14. Prove that if in an abelian group the element a has order k and the element b has order l, and if k and l are relatively prime, then the element ab has order kl. [Hint: If $(ab)^t = e$, raise both sides to the power k and conclude that t must be divisible by l. Similarly, show that t must be divisible by k.]

7.5 COSETS AND LAGRANGE'S THEOREM

Let G be an arbitrary group and H a subgroup of G. If $a \in G$, we shall designate by aH the *set* of all elements of G of the form ah, where $h \in H$. That is, $aH = \{ah \mid h \in H\}$.

7.32 Definition. If H is a subgroup of the group G and $a \in G$, we call aH a *coset* of H (in G).*

Since $eH = H$, we see that H is itself a coset. Moreover, since $e \in H$, it is clear that $a \in aH$.

The following lemma will be very useful in studying cosets.

7.33 Lemma. *If H is a subgroup of the group G and $a, b \in G$, then each of the following is true:*

(i) If $aH \cap bH \neq \varnothing$, then $aH = bH$.
(ii) $aH = bH$ if and only if $a \in bH$.

PROOF OF (i). Suppose that aH and bH have at least one element in common. Thus, there exist $h_1, h_2 \in H$ such that $ah_1 = bh_2$. Then $a = bh_2h_1^{-1}$ and any element ah of aH can be expressed in the form $bh_2h_1^{-1}h$. Since $h_2h_1^{-1}h \in H$, it follows that $ah \in bH$. We have therefore shown that $aH \subseteq bH$. In a similar way we can show that $bH \subseteq aH$ and therefore we conclude that $aH = bH$. One way

* More precisely, we have here defined a *left* coset, and one can similarly define a right coset Ha. However, in accordance with the definition just given we shall in this section use the word *coset* to mean *left coset*.

of stating the property we have just proved is to say that two cosets either coincide or have no element in common.

Since $a \in aH$, each element of G is in some coset; thus the property just proved shows that the different cosets of H in G form a partition of G.

PROOF OF (ii). Since $a \in aH$ it is obvious that if $aH = bH$, then $a \in bH$. Conversely, suppose that $a \in bH$. Then $a \in aH \cap bH$, and part (i) of the lemma implies at once that $aH = bH$.

As an example of cosets, consider the symmetric group S_3 with multiplication table 7.13. We know that α_1 is the identity of this group, and it is easy to verify that $H = \{\alpha_1, \alpha_2\}$ is a subgroup. By use of the table, we find the following cosets of H in S_3:

$$\begin{aligned} \alpha_1 H &= \{\alpha_1, \alpha_2\}, & \alpha_4 H &= \{\alpha_4, \alpha_5\}, \\ \alpha_2 H &= \{\alpha_2, \alpha_1\}, & \alpha_5 H &= \{\alpha_5, \alpha_4\}, \\ \alpha_3 H &= \{\alpha_3, \alpha_6\}, & \alpha_6 H &= \{\alpha_6, \alpha_3\}. \end{aligned}$$

We see, therefore, that there are three different cosets of H in S_3, that every coset contains two elements, and that every element of S_3 is in exactly one of these three cosets. As an illustration of Lemma 7.33(ii), we may observe that $\alpha_6 H = \alpha_3 H$ since $\alpha_6 \in \alpha_3 H$, but that $\alpha_4 H \neq \alpha_3 H$ since $\alpha_4 \notin \alpha_3 H$.

So far we have used multiplication as the operation but, as usual, it is easy to make the necessary modifications if the operation is addition. In this case, a coset is of the form $a + H = \{a + h \mid h \in H\}$. As an example, let G be the additive group of the ring $\mathbf{Z}_{12}$ and H the subgroup $\{[0], [3], [6], [9]\}$. Then it may be verified that the *different* cosets of H in G are the following:

$$\begin{aligned} [0] + H &= \{[0], [3], [6], [9]\}, \\ [1] + H &= \{[1], [4], [7], [10]\}, \\ [2] + H &= \{[2], [5], [8], [11]\}. \end{aligned}$$

Let us now make the following definition.

7.34 Definition. If the group G has finite order and H is a subgroup of G, the number of distinct cosets of H in G is called the *index* of H in G.

Although we have been using *coset* to mean *left coset*, we should perhaps point out that Exercise 7 below shows that there are the same number of right cosets as of left cosets of H in G. Accordingly, in the definition just given it does not matter whether we think of left cosets or of right cosets.

We shall next prove the following theorem of Lagrange which is of fundamental importance in the study of groups of finite order.

7.35 Theorem. *Suppose that the group G has order n. If H is a subgroup of G of order m and of index k, then $n = km$. In particular, both the order and the index of H are divisors of the order of G.*

PROOF. We first observe that every coset of H in G has exactly m elements. For if $a \in G$ and $h_1, h_2 \in H$, then $ah_1 = ah_2$ if and only if $h_1 = h_2$. Hence an arbitrary coset aH has the same number of elements as H, namely, m.

We have already shown that the distinct cosets of H in G form a partition of G. Since there are k distinct cosets and each of them contains m elements, G must contain km elements. This shows that $n = km$ and the proof is complete.

There are some interesting consequences of the fact that the order of a subgroup of a finite group is a divisor of the order of the group. First of all, since the order of an element of a group is the order of the cyclic subgroup generated by that element, we have at once the following corollary.

7.36 Corollary. *The order of an element of a group of finite order is a divisor of the order of the group.*

If the order of a group is a prime p, then every element of the group, other than the identity, must have order p. This yields the next result as follows.

7.37 Corollary. *A group of order p, where p is a prime, is a cyclic group. Moreover, every element except the identity is a generator of the group.*

If the group G has order n, and the element a of G has order m, then, by Corollary 7.36, we have $n = mk$ for some integer k. By Theorem 7.26, we know that $a^m = e$, and hence $a^n = (a^m)^k = e^k = e$. We have established the following corollary.

7.38 Corollary. *If a is an element of a group of order n, then $a^n = e$.*

If the integer s is not divisible by the prime p, then $[s]$ is an element of the multiplicative group of the field $\mathbf{Z}_p$. Since this group has order $p - 1$, the preceding corollary shows that $[s]^{p-1} = [1]$. However, this implies that $[s^{p-1}] = [1]$ and we conclude that $s^{p-1} \equiv 1 \pmod{p}$. We have therefore obtained the following theorem of Fermat.

7.39 Corollary. *If s is an integer not divisible by the prime p, then $s^{p-1} \equiv 1 \pmod{p}$.*

EXERCISES

1. Exhibit all cosets of the subgroup $\{\epsilon, \alpha, \alpha^2, \alpha^3\}$ of the octic group (7.15).
2. Let G be the multiplicative group of the field $\mathbf{Z}_{19}$ and H the cyclic subgroup of G generated by the element $[8]$. Exhibit all of the cosets of H in G.
3. Exhibit all cosets of the subgroup $\{0, 4, 8, 12, 16\}$ of the additive group of $\mathbf{Z}_{20}$.
4. Prove that a group of order n has a proper subgroup if and only if n is not a prime.
5. Prove that if an abelian group G of order 6 contains an element of order 3, G must be a cyclic group.
6. Let H be a subgroup of a group G. If $a, b \in G$, let $a \sim b$ mean that $b^{-1}a \in H$. Show that "$\sim$" is an equivalence relation defined on G. If $[a]$ is the equivalence set which contains a, show that $[a] = aH$ and therefore the cosets of H in G are the equivalence sets relative to this equivalence relation.
7. Let H be a subgroup of a group G and define a mapping β of the set of left cosets of H into the set of right cosets of H as follows: $(aH)\beta = Ha^{-1}$. Prove that β is a well-defined mapping and that, in fact, it is a one-one mapping of the set of all left cosets of H onto the set of all right cosets.
8. Let G be a group of order n, and let H and K be subgroups of G of respective orders h and k. If the subgroup $D = H \cap K$ of G has order d and we set $HK = \{hk \mid h \in H, k \in K\}$, prove that HK contains exactly hk/d distinct elements. [Hint: If $\{Dk_1, Dk_2, \cdots, Dk_m\}$ is a complete set of distinct cosets of D in K, show that $\{Hk_1, Hk_2, \cdots, Hk_m\}$ is a partition of HK.]
9. Let H and K be subgroups of a group G. If $a \in G$, the set $HaK = \{hak \mid h \in H, k \in K\}$ is called a *double coset* of H and K in G.
   (i) If $a, b \in G$ and $HaK \cap HbK \neq \varnothing$, prove that $HaK = HbK$.
   (ii) If G has finite order, either prove that all double cosets of fixed subgroups H and K in G have the same number of elements or give an example to show that this need not be true.

7.6 THE SYMMETRIC GROUP S_n

We now return to a further study of permutations of a finite set $A = \{1, 2, \cdots, n\}$. We have already defined the symmetric group S_n to be the group of all permutations of A. Throughout this section the word *permutation* will mean an element of S_n for some positive integer n, and we shall sometimes find it convenient to refer to the elements of A as "symbols."

We shall first study permutations of the particular type described in the following definition.

7.40 Definition. An element α of S_n is said to be a *cycle of length k* if there exist distinct elements $a_1, a_2, \cdots, a_k$ $(k \geq 1)$ of A such that

$$a_1\alpha = a_2, \quad a_2\alpha = a_3, \quad \cdots, \quad a_{k-1}\alpha = a_k, \quad a_k\alpha = a_1,$$

and $i\alpha = i$ for each element i of A other than $a_1, a_2, \cdots, a_k$. This cycle α may be designated by $(a_1a_2 \cdots a_k)$.

It will be observed that a cycle of length 1 is necessarily the identity permutation. It sometimes simplifies statements to consider the identity permutation as a cycle, but we shall usually be interested in cycles of length greater than 1.

As an example of a cycle, suppose that β is the element of S_6 defined by

$$1\beta = 3, \quad 3\beta = 2, \quad 2\beta = 5, \quad 5\beta = 6, \quad 6\beta = 1, \quad 4\beta = 4.$$

Then β is a cycle of length 5, and we may write $\beta = (13256)$. In a cycle, such as (13256), the symbols appearing are permuted cyclically; that is, each symbol written down maps into the next one, except that the last maps into the first. A symbol, such as 4 in this example, which is not written down is assumed to map into itself. There are other ways of writing the cycle defined above. For example, $\beta = (32561) = (25613)$, and so on. Also, in another notation introduced in Section 7.3, we have

$$\beta = \begin{pmatrix} 1 & 2 & 3 & 4 & 5 & 6 \\ 3 & 5 & 2 & 4 & 6 & 1 \end{pmatrix}.$$

As further illustrations of all the various notations used, let us consider elements of S_6 and verify that

$$(1345)(146) = \begin{pmatrix} 1 & 2 & 3 & 4 & 5 & 6 \\ 3 & 2 & 6 & 5 & 4 & 1 \end{pmatrix}.$$

In the first factor 1 maps into 3, and in the second factor 3 is unchanged; hence in the product, 1 maps into 3. The symbol 2 does not appear in either factor; hence 2 maps into 2. In the left factor 3 maps into 4, and then in the second factor 4 maps into 6; hence in the product, 3 maps into 6. Similarly, the other verifications are easily made.

Now let α be the cycle $(a_1a_2 \cdots a_k)$ of S_n of length k, and let us consider the powers of α. Under the mapping α^2, we see that a_1 maps into a_3 (if $k \geq 3$), for

$$a_1\alpha^2 = (a_1\alpha)\alpha = a_2\alpha = a_3 .$$

Similarly, under the mapping α^3, a_1 maps into a_4 (if $k \geq 4$), and so on. Continuing, we find that $a_1\alpha^k = a_1$. Since we could just as well write $\alpha = (a_2a_3 \cdots a_ka_1)$, a similar argument shows that $a_2\alpha_k = a_2$ and, in general, that $a_i\alpha^k = a_i$ for $i = 1, 2, \cdots, k$. It follows that $\alpha_k = \epsilon$, the identity permutation, and, moreover, k is the smallest power of α which is equal to ϵ. The following result then follows immediately from Theorem 7.26.

7.41 Theorem. *A cycle of length k has order k.*

Two cycles $(a_1a_2 \cdots a_k)$ and $(b_1b_2 \cdots b_l)$ of S_n are said to be *disjoint* if the sets $\{a_1, a_2, \cdots, a_k\}$ and $\{b_1, b_2, \cdots, b_l\}$ have no elements in common. A set of more than two cycles is said to be disjoint if each pair of them is disjoint. The next result shows why cycles play an important role in the study of permutations.

7.42 Theorem. *Every element γ of S_n that is not itself a cycle is expressible as a product of disjoint cycles.*

Before considering the proof, let us look at an example. Suppose that

$$\gamma = \begin{pmatrix} 1 & 2 & 3 & 4 & 5 & 6 \\ 3 & 1 & 4 & 2 & 6 & 5 \end{pmatrix},$$

and let us start with any symbol which does not map into itself, for example, the symbol 1. We see that $1\gamma = 3$, $3\gamma = 4$, $4\gamma = 2$, and $2\gamma = 1$. Now take any symbol which has not yet been used and which does not map into itself, for example 5. Then $5\gamma = 6$, and $6\gamma = 5$. It is then almost obvious that $\gamma = (1342)(56)$.

PROOF. The proof in the general case follows the same pattern as in this example. Since the identity permutation is a cycle (of length 1), we assume that γ is not the identity. Start with any symbol a_1 such that $a_1\gamma \neq a_1$, and suppose that $a_1\gamma = a_2$, $a_2\gamma = a_3$, $a_3\gamma = a_4$, and so on until we come to the point where, say, $a_k\gamma$ equals some one of the symbols $a_1, a_2, \cdots, a_{k-1}$ already used. Then we must have $a_k\gamma = a_1$ since every other one of these symbols is already known to be the image of some symbol under the mapping γ. Thus γ has the same effect on the symbols $a_1, a_2, \cdots, a_k$ as the cycle $(a_1a_2 \cdots a_k)$, and also effects a permutation of the remaining symbols (if any). If b_1 is a symbol other than $a_1, a_2, \cdots, a_k$ and $b_1\gamma \neq b_1$, we proceed as above and obtain a cycle $(b_1b_2 \cdots b_l)$. Now if all symbols that do not map into themselves have been used, we have

$$\gamma = (a_1a_2 \cdots a_k)(b_1b_2 \cdots b_l).$$

If there is another symbol c_1 such that $c_1\gamma \neq c_1$, we can similarly obtain another cycle. Evidently, the process can be continued to obtain the desired result. A complete proof can be given by induction.

EXERCISES

1. In each of the following, γ is an element of S_7. Express it as a product of disjoint cycles.

(a) $1\gamma = 3$, $2\gamma = 4$, $3\gamma = 1$, $4\gamma = 7$, $5\gamma = 5$, $6\gamma = 6$, $7\gamma = 2$.
(b) $1\gamma = 5$, $2\gamma = 3$, $3\gamma = 4$, $4\gamma = 7$, $5\gamma = 6$, $6\gamma = 1$, $7\gamma = 2$.

(c) $\gamma = \begin{pmatrix} 1 & 2 & 3 & 4 & 5 & 6 & 7 \\ 3 & 4 & 1 & 2 & 6 & 7 & 5 \end{pmatrix}$.

(d) $\gamma = \begin{pmatrix} 1 & 2 & 3 & 4 & 5 & 6 & 7 \\ 2 & 3 & 1 & 5 & 4 & 7 & 6 \end{pmatrix}$.

2. Express each of the following elements of S_7 as a product of disjoint cycles:

(a) (123)(16543),
(b) (213456)(172),
(c) (4215)(3426)(5671),
(d) (1234)(124)(3127)(56).

The cycles of length 2 are of special interest, and we make the following definition.

7.43 Definition. A cycle of length 2 is called a *transposition.*

A transposition (ij) merely interchanges the symbols i and j, and leaves the other symbols unchanged. Since $(ij)(ij) = \epsilon$, it follows that *a transposition is its own inverse.*

It is quite easy to show that every cycle of length more than 2 can be expressed as a product of transpositions. In fact, this result follows from the observation that

$$(a_1a_2 \cdots a_k) = (a_1a_k)(a_2a_k) \cdots (a_{k-1}a_k),$$

which can be verified by direct calculation. In view of Theorem 7.42, it follows immediately that *every* permutation can be expressed as a product of transpositions. However, it is easy to verify that there is more than one way to express a permutation as such a product. As examples, we see that

$$\begin{aligned}(1234) &= (14)(24)(34) = (32)(12)(14) = (13)(24)(34)(12)(24),\\ (123)(14) &= (12)(13)(14) = (14)(24)(34) = (14)(24)(34)(23)(23),\end{aligned}$$

and so on. Since $(ij)(ij) = \epsilon$, we can insert as many such pairs of identical transpositions as we wish. Clearly, then, a permutation can be expressed as a product of transpositions in many different ways.

The following theorem, of which the first statement has already been proved, is one of the principal theorems about permutations.

7.44 Theorem. *Every permutation α can be expressed as a product of transpositions. Moreover, if α can be expressed as a product of r transpositions and also as a product of s transpositions, then either r and s are both even or they are both odd.*

PROOF. Suppose that α is a permutation of the set $A = \{1, 2, \cdots, n\}$. Suppose, further, that

7.45 $$\alpha = \beta_1\beta_2 \cdots \beta_r = \gamma_1\gamma_2 \cdots \gamma_s,$$

where each β and each γ is a transposition. To establish the theorem, we need to prove that r and s are both even or that they are both odd. There are ways to prove this fact by calculating entirely with permutations, but we proceed to give a well-known proof which is simpler in its details but which involves the introduction of a certain "counting device" which has no inherent connection with the permutations themselves. Let $x_1, x_2, \cdots, x_n$

be independent symbols (or variables, if you wish) and let P denote the polynomial* with integral coefficients defined as follows:

7.46
$$P = \prod_{i<j} (x_i - x_j),$$

it being understood that this stands for the product of all expressions of the form $x_i - x_j$, where i and j take values from 1 to n, with $i < j$. We now define

7.47
$$P\alpha = \prod_{i<j} (x_{i\alpha} - x_{j\alpha}),$$

that is, $P\alpha$ is the polynomial obtained by performing the permutation α on the *subscripts* of the symbols $x_1, x_2, \cdots, x_n$.

As an illustration of this notation, if $n = 4$, we have

$$P = (x_1 - x_2)(x_1 - x_3)(x_1 - x_4)(x_2 - x_3)(x_2 - x_4)(x_3 - x_4).$$

Moreover, if

$$\alpha = \begin{pmatrix} 1 & 2 & 3 & 4 \\ 4 & 1 & 2 & 3 \end{pmatrix},$$

we find that

$$P\alpha = (x_4 - x_1)(x_4 - x_2)(x_4 - x_3)(x_1 - x_2)(x_1 - x_3)(x_2 - x_3),$$

and it is easily verified that $P\alpha = -P$. In general, it is fairly clear that always $P\alpha = \pm P$, with the sign depending in some way on the permutation α.

We next prove the following lemma.

7.48 Lemma. *If $\delta = (kl)$ is a transposition, then $P\delta = -P$.*

PROOF. Now $k \neq l$, and there is no loss of generality in assuming that $k < l$. Hence, one of the factors in P is $x_k - x_l$ and in $P\delta$ the corresponding factor is $x_l - x_k$; that is, this factor is just changed in sign under the mapping δ on the subscripts. Any factor of P of the form $x_i - x_j$, where neither i nor j is equal to k or l, is clearly unchanged under the mapping δ. All other factors of P can be paired to form products of the form $\pm(x_i - x_k)(x_i - x_l)$, with the sign determined by the relative magnitudes of i, k, and l.

* Logically, this proof should be deferred until after polynomials have been studied in some detail. However, it seems preferable to insert it here with the expectation that it will be convincing, even though use is made of a few simple properties of polynomials which will not be established until Chapter 9.

But since the effect of δ is just to interchange x_k and x_l, any such product is unchanged. Hence, the only effect of δ is to change the sign of P, and the lemma is established.

The proof of the theorem now follows easily. Since, by 7.45, $P\alpha$ can be computed by performing in turn the r transpositions $\beta_1, \beta_2, \cdots, \beta_r$, and by the lemma each of these merely changes the sign of P, it follows that $P\alpha = (-1)^r P$. In like manner, using the fact that $\alpha = \gamma_1\gamma_2 \cdots \gamma_s$, we see that also $P\alpha = (-1)^s P$. Hence, we must have $(-1)^r P = (-1)^s P$, from which it follows that $(-1)^r = (-1)^s$. This implies that r and s are both even or they are both odd, and the proof is complete.

7.49 Definition. A permutation is called an *even* permutation or an *odd* permutation according as it can be expressed as a product of an even or an odd number of transpositions.

If the permutation α can be expressed as a product of k transpositions and the permutation β can be expressed as a product of l transpositions, it is obvious that $\alpha\beta$ can be expressed as a product of $k + l$ transpositions. It follows that the product of two even, or of two odd, permutations is an even permutation, whereas the product of an odd permutation and an even permutation is an odd permutation.

Another observation of some importance is the following. Suppose that α is a product of k transpositions, say $\alpha = \alpha_1\alpha_2 \cdots \alpha_k$. Then, since a transposition is its own inverse, it is easy to see that $\alpha^{-1} = \alpha_k\alpha_{k-1} \cdots \alpha_1$. It follows that α^{-1} is an even permutation if and only if α is an even permutation. We shall conclude our study of permutation groups by proving the following theorem.

7.50 Theorem. *The set A_n of all even permutations of the symmetric group S_n is a subgroup of S_n of order $n!/2$.*

PROOF. The fact that A_n is a subgroup of S_n follows at once from the preceding remarks and Theorem 7.4. This subgroup A_n of S_n is usually called the *alternating group* on n symbols.

Let us now consider the order of A_n. If β is a fixed odd permutation, all the elements of the coset βA_n are odd permutations since the product of an odd permutation by an even permutation is necessarily an odd permutation. We proceed to show that *all* odd permutations of S_n are in the coset βA_n. If γ is an arbitrary odd permutation, we may write $\gamma = \beta(\beta^{-1}\gamma)$, and $\beta^{-1}\gamma$ is an even permutation since β^{-1} and γ are both odd. It follows that $\beta^{-1}\gamma \in A_n$, and hence that $\gamma \in \beta A_n$. We have shown that the coset βA_n consists of *all* the odd permutations, and hence that there are just the two cosets A_n and βA_n of A_n in S_n. Since these cosets have the same number of elements and S_n has order $n!$, it follows that the alternating group A_n has order $n!/2$.

EXERCISES

1. Verify that a cycle of length k is an even or an odd permutation according as k is odd or even, respectively.
2. Prove that every even permutation is a cycle of length 3 or can be expressed as a product of cycles of length 3. [Hint: $(12)(13) = (123)$, and $(12)(34) = (134)(321)$.]
3. Exhibit the elements of the alternating group A_3 and of the alternating group A_4.
4. Let G be a subgroup of the symmetric group S_n, and suppose that G contains at least one odd permutation. By a suitable modification of the proof of Theorem 7.50, prove that the set of all even permutations in G is a subgroup of G, and then prove that G contains the same number of odd permutations as of even permutations.

7.7 NORMAL SUBGROUPS AND QUOTIENT GROUPS

If H is a subgroup of the group G and $a \in G$, the set $aH = \{ah \mid h \in H\}$, which in Section 7.5 was called a coset, we shall for the present call a *left* coset of H in G. Similarly, the set $Ha = \{ha \mid h \in H\}$ is called a *right* coset of H in G. It need not be true that a left coset aH is equal to the right coset Ha. However, we shall be interested in subgroups which do have this property, and we therefore introduce the following definition.

7.51 Definition. A subgroup K of a group G is said to be a *normal* (or *invariant*) subgroup of G if and only if $aK = Ka$ for every element a of G.

We may emphasize that this definition does not state that necessarily $ak = ka$ for each $a \in G$ and $k \in K$; it merely states that the *sets* aK and Ka coincide. In particular, if $a \in G$ and $k \in K$, there must exist an element k_1 of K (not necessarily the same element k) such that $ak = k_1a$.

Clearly, every subgroup of an abelian group is a normal subgroup. As an example of a subgroup which is not normal, consider the symmetric group S_3 with multiplication table 7.13 and the subgroup $H = \{\alpha_1, \alpha_2\}$ of S_3. It may be verified that $\alpha_3 H = \{\alpha_3, \alpha_6\}$, whereas $H\alpha_3 = \{\alpha_3, \alpha_5\}$, and hence that $\alpha_3 H \neq H\alpha_3$, so H is not a normal subgroup of S_3. However, S_3 has as normal subgroup the alternating group $A_3 = \{\alpha_1, \alpha_5, \alpha_6\}$. Instead of verifying this fact by direct calculation, let us prove that for each positive integer $n > 1$, *the alternating group* A_n *is a normal subgroup of the symmetric group* S_n. If α is an even permutation, then $\alpha \in A_n$ and $\alpha A_n = A_n\alpha = A_n$. If α is an odd permutation, the proof of Theorem 7.50 shows that αA_n is the set of all odd permutations of S_n. A similar argument will show that also $A_n\alpha$ is the set of all odd permutations of S_n. Hence for every $\alpha \in S_n$ we have $\alpha A_n = A_n\alpha$, and A_n is therefore a normal subgroup of S_n.

Now let G be an arbitrary group and K a normal subgroup of G. Since K is normal, we need not distinguish between left cosets and right cosets; so we shall again simply call them cosets and write them as left cosets. On the set of all cosets of K in G we propose to define an operation of multiplication which will make this set into a group. Accordingly, let us define

7.52 $$(aK)(bK) = (ab)K, \qquad a, b \in G.$$

In order to verify that this does define an operation on the set of all cosets, we need to show that multiplication is well-defined by this equation. That is, we must show that if $aK = a_1K$ and $bK = b_1K$, then $(ab)K = (a_1b_1)K$. By 7.33(ii), this fact can be established by showing that if $a \in a_1K$ and $b \in b_1K$, then $ab \in (a_1b_1)K$. Suppose then, that $a = a_1k$ and $b = b_1k'$, where $k, k' \in K$. Thus $ab = a_1kb_1k'$ and, since K is a normal subgroup of G, there exists $k'' \in K$ such that $kb_1 = b_1k''$. Hence $ab = a_1b_1k''k'$ and it follows that $ab \in (a_1b_1)K$, as we wished to show. This proves that 7.52 does indeed define an operation of multiplication on the set of all cosets of K in G, and we proceed to prove the following theorem.

7.53 Theorem. *Let K be a normal subgroup of the group G. With respect to the multiplication 7.52 of cosets, the set of all cosets of K in G is a group, usually called the* quotient group *of G by K and denoted by G/K. Moreover, the mapping $\theta\colon G \to G/K$ defined by $a\theta = aK$, $a \in G$, is a homomorphism of G onto G/K, with kernel K.*

PROOF. The associative law in G/K is an almost immediate consequence of the associative law in G, and we leave this part of the proof to the reader. Now, if e is the identity of G, then since by 7.52,

$$(aK)(eK) = (eK)(aK) = aK, \qquad a \in G,$$

we see that $eK = K$ is the identity of G/K. Finally, 7.52 implies that

$$(aK)(a^{-1}K) = (aa^{-1})K = eK = K$$

and, similarly, $(a^{-1}K)(aK) = K$. Hence $a^{-1}K$ is the inverse of aK, and we have proved that G/K is a group. (Cf. Exercise 14 of Section 7.3.) Furthermore, the definition of the mapping θ shows that it is a mapping of G onto G/K, and the definition of multiplication of cosets shows that

$$(ab)\theta = (ab)K = (aK)(bK) = (a\theta)(b\theta), \qquad a, b \in G,$$

and hence that θ is a homomorphism. Finally, since K is the identity of G/K, an element a of G is in ker θ if and only if $aK = K$, that is, if and only if $a \in K$. This completes the proof of the theorem.

We may point out that since the elements of the quotient group G/K are the distinct cosets of K in G, if G has finite order, the order of the group G/K is the index of K in G. In fact, if G has finite order, Theorem 7.35 shows that

$$\text{order of } G/K = \frac{\text{order of } G}{\text{order of } K}.$$

In the above, we have used multiplication as the operation in G. If G is abelian and the operation is considered to be addition, it is important to keep in mind that a coset is of the form $a + K$, and the multiplication 7.52 of cosets is replaced by addition of cosets defined as follows:

7.54 $$(a + K) + (b + K) = (a + b) + K, \qquad a, b \in G.$$

In this case, the identity of G is called the "zero" as usual, and the zero of the quotient group G/K is the coset K.

We have shown that if K is a normal subgroup of an arbitrary group G, then there exists a homomorphism of G, with kernel K, onto the quotient group G/K. We shall next prove that "essentially" all homomorphisms of G are of this type. More precisely, we shall show that the kernel of every homomorphism of G is a normal subgroup of G and that every homomorphic image of G is isomorphic to a quotient group G/K for some choice of the normal subgroup K. This is the content of the following theorem.

7.55 Fundamental Theorem on Group Homomorphisms. *Let $\phi: G \to H$ be a homomorphism of the group G onto the group H with kernel K. Then K is a normal subgroup of G, and H is isomorphic to the quotient group G/K. More precisely, the mapping $\alpha: G/K \to H$ defined by*

7.56 $$(aK)\alpha = a\phi, \qquad a \in G,$$

is an isomorphism of G/K onto H.

PROOF. Let us first show that the kernel K of a homomorphism $\phi: G \to H$ is necessarily a normal subgroup of G. We have already proved in Theorem 7.21 that K must be a subgroup, so there only remains to prove that it is normal. If $a \in G$ and $k \in K$, then

$$(aka^{-1})\phi = (a\phi)(k\phi)(a\phi)^{-1}.$$

But if e is the identity of G, $e\phi$ is the identity of H and $k\phi = e\phi$ by definition of ker ϕ. Thus $(aka^{-1})\phi = e\phi$ and $aka^{-1} \in \ker \phi = K$. Hence $aka^{-1} = k_1$ for some element k_1 of K. It follows that $ak = k_1a$ and this shows that $aK \subseteq Ka$. In like manner, it can be shown that $Ka \subseteq aK$, so that $aK = Ka$ and K is indeed a normal subgroup. Thus we can now speak of the quotient group G/K.

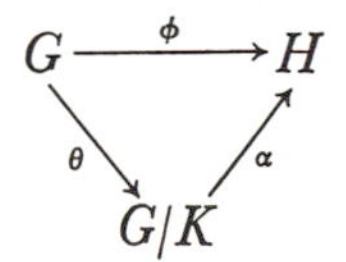

Before proving the rest of the theorem, its meaning may perhaps be clarified by reference to the accompanying diagram. Here ϕ is the given homomorphism of G onto H, $K = \ker \phi$, and $\theta: G \to G/K$ is the homomorphism of G onto G/K defined by $a\theta = aK$, $a \in G$, as in the preceding theorem. Our present theorem may then be interpreted as stating that $\phi = \theta\alpha$, that is, that $a\phi = (a\theta)\alpha = (aK)\alpha$ for each $a \in G$. Otherwise expressed, an element a of G has the same image in H no matter which of the two paths from G to H is taken.

Let us now show that a mapping α of G/K into H is well-defined by 7.56. That is, we shall show that if $aK = a_1K$, then $a\phi = a_1\phi$. If $aK = a_1K$, we have that $a = a_1k$ for some element k

of K. Hence $a\phi = (a_1\phi)(k\phi) = a_1\phi$ since $k \in \ker \phi$ and therefore $k\phi$ is the identity of H. This proves that the mapping α is well-defined.

Next, using the fact that ϕ is a homomorphism, and the definition of multiplication of cosets, we see that for $a, b \in G$,

$$[(aK)(bK)]\alpha = [(ab)K]\alpha = (ab)\phi = (a\phi)(b\phi) = [(aK)\alpha][(bK)\alpha].$$

This shows that α is a homomorphism and it is clearly a mapping onto H. There remains only to prove that it is an isomorphism.

Suppose that $aK \in \ker \alpha$. Hence, if e is the identity of G, and therefore $e\phi$ is the identity of H, we have $(aK)\alpha = e\phi$ or $a\phi = e\phi$. However, this implies that $a \in \ker \phi = K$, so $aK \in \ker \alpha$ implies that $a \in K$ and therefore that $aK = K$. Since K is the identity of the group G/K, we conclude that $\ker \alpha$ consists only of the identity of G/K. By Theorem 7.21 it follows that α is an isomorphism, and the theorem is proved.

EXERCISES

1. If H is a subgroup of a group G and $a \in G$, prove that $aHa^{-1} = \{aha^{-1} \mid h \in H\}$ is a subgroup of G which is isomorphic to H.
2. Prove that the intersection of two or more normal subgroups of a group G is a normal subgroup of G.
3. If G is a group, prove that the set $\{a \mid a \in G, ax = xa \text{ for every } x \in G\}$ is a normal subgroup of G.
4. Verify that the subgroup $\{\epsilon, \alpha, \alpha^2, \alpha^3\}$ of the octic group (7.15) is a normal subgroup.
5. If e is the identity of G, and H and K are normal subgroups of G with $H \cap K = \{e\}$, prove that $hk = kh$ for any $h \in H$, $k \in K$. [Hint: Show that $h^{-1}k^{-1}hk \in H \cap K$.]
6. Prove that if every right coset of a subgroup H in G is also a left coset, then H is necessarily normal. [Hint: If $Ha = bH$, show that $bH = aH$.]
7. Let G be the cyclic group of order 20 generated by an element a, and let H be the subgroup of G generated by a^4. Write out the cosets of H in G and verify that the quotient group G/H is a cyclic group of order 4.
8. It can be shown that the set of elements $K = \{(1), (12)(34), (13)(24), (14)(23)\}$ of the alternating group A_4 on four symbols is a normal subgroup of A_4. Without calculation, explain how you know that the quotient group A_4/K must be cyclic.
9. Prove: If there exist exactly two left cosets (or right cosets) of a subgroup H in a group G, then H is necessarily a normal subgroup of G.

10. Let **Q** be the additive group of the field of rational numbers and **Z** the additive group of the ring of integers. Show that every element of the quotient group **Q**/**Z** has finite order. Does the group **Q**/**Z** have finite order?

11. Let K be a normal subgroup of the group G, and let A be a subgroup of the quotient group G/K. Thus we may consider A to be a set of cosets of K in G. Prove that the union of these cosets is a subgroup of G.

12. Let H and K be subgroups of a group G, with K a normal subgroup of G. Prove each of the following:
(i) $H \cap K$ is a normal subgroup of H.
(ii) If $HK = \{hk \mid h \in H, k \in K\}$, then HK is a subgroup of G.
(iii) K is a normal subgroup of the group HK.

13. If H and K are as in the preceding exercise, prove each of the following:
(i) Every element of the quotient group HK/K is expressible in the form hK, $h \in H$.
(ii) The mapping $\alpha: H \to HK/K$ defined by $h\alpha = hK$, $h \in H$, is a homomorphism of H onto HK/K, with kernel $H \cap K$.
(iii) The group $H/(H \cap K)$ is isomorphic to the group HK/K.

14. Let $\theta: G \to H$ be a homomorphism of the group G onto the group H, with kernel $\theta = K$. If A is a subset of G, let us use the notation introduced in Section 1.2 and write $A\theta$ for the set of elements of H which occur as images of elements of A under the mapping θ. In particular, $G\theta = H$ since θ is given as an onto mapping. If U is a subset of H, let us define $U\theta^{-1} = \{x \mid x \in G, x\theta \in U\}$.* Prove each of the following
(i) If A is a subgroup of G, then $A\theta$ is a subgroup of H.
(ii) If U is a subgroup of H, then $U\theta^{-1}$ is a subgroup of G which contains K.
(iii) The mapping $A \to A\theta$ is a one-one mapping of the set of all subgroups of A which contain K onto the set of all subgroups of H.

NOTES AND REFERENCES

Items [14] through [17] of the Bibliography is a short list of books on the theory of groups. Lederman [16] is particularly recommended as giving a simple introduction to parts of the subject. Hall [15] and Rotman [17] are quite extensive, and the latter is particularly modern in approach. Baumslag and Chandler [14] is really an extensive outline of the subject and contains the solution of many problems. Also, many of the books on abstract algebra in general devote a considerable amount of space to group theory.

* This is here a convenient notation but is not to be confused with the inverse mapping as introduced in Section 1.2. Actually, as here used, θ^{-1} is not a mapping of H into G, but it does define a mapping of the set of all subsets of H into the set of all subsets of G.

Chapter 8

Finite Abelian Groups

The problem of defining all finite groups is a difficult and, in fact, an unsolved problem. However, in a sense to be made precise later, it is possible to determine all finite *abelian* groups. The purpose of this chapter is to prove the fundamental results in the theory of such groups.

Let G be a finite abelian group. Throughout this chapter we shall use addition as the operation in G. Of course, everything could just as well be stated in terms of multiplication as the operation. Unless otherwise explicitly stated, we shall always assume that the group G under discussion is a nonzero group, that is, that it does not consist of the identity alone.

Let us recall the following essential facts which will be used frequently in the sequel. Since G has finite order, each element a of G has finite order. If a has order n, Lemma 7.27 shows that n is the least positive integer such that $na = 0$. The order of the zero element is one, all other elements have order greater than one. If a has order n and $k \in \mathbf{Z}$, then $ka = 0$ if and only if $n|k$.

8.1 DIRECT SUMS OF SUBGROUPS

If $G_1, G_2, \cdots, G_r$ are subgroups of the abelian group G, we define the *sum*

$$G_1 + G_2 + \cdots + G_r$$

of these subgroups to be the set of all elements of G which can be expressed in the form

$$a_1 + a_2 + \cdots + a_r, \qquad a_i \in G_i \ (i = 1, 2, \cdots, r).$$

This set is seen to be a subgroup of G, and each G_i is contained in this subgroup since the identity 0 of G is an element of each G_i. Actually, this sum is the smallest subgroup of G which contains all the subgroups G_i.

We now make the following definition.

8.1 Definition. If G_i $(i = 1, 2, \cdots, r)$ are subgroups of the abelian group G, the sum $G_1 + G_2 + \cdots + G_r$ is said to be a *direct sum* if and only if the following condition is satisfied:

(i) If $a_i \in G_i$ $(i = 1, 2, \cdots, r)$ such that

$$a_1 + a_2 + \cdots + a_r = 0,$$

then each $a_i = 0$.

We shall indicate that a sum $G_1 + G_2 + \cdots + G_r$ is a direct sum by writing it in the form

$$G_1 \oplus G_2 \oplus \cdots \oplus G_r.$$

It is worth pointing out that condition (i) is equivalent to the following condition:

(ii) If $a_i, b_i \in G$ $(i = 1, 2, \cdots, r)$ such that

$$a_1 + a_2 + \cdots + a_r = b_1 + b_2 + \cdots + b_r,$$

then $a_i = b_i$ $(i = 1, 2, \cdots, r)$.

The equivalence of (i) and (ii) follows readily from the fact that the equation $a_1 + a_2 + \cdots + a_r = b_1 + b_2 + \cdots + b_r$ may be written in the form

$$(a_1 - b_1) + (a_2 - b_2) + \cdots + (a_r - b_r) = 0.$$

We leave the details of the proof of the equivalence of conditions (i) and (ii) as an exercise.

The condition (ii) for a sum $G_1 + G_2 + \cdots + G_r$ to be a direct sum is often expressed by saying that the sum is direct if and only if each element of the sum is *uniquely* expressible in the form

$$a_1 + a_2 + \cdots + a_r, \qquad a_i \in G_i \ (i = 1, 2, \cdots, r).$$

If G_i has order n_i, we see that in a sum of the type just written there are n_i choices for a_i, and the uniqueness property just mentioned shows that the order of a direct sum $G_1 \oplus G_2 \oplus \cdots \oplus G_r$ is the product $n_1 n_2 \cdots n_r$ of the orders of the respective subgroups G_i.

Since addition is a commutative operation in G, it is clear, for example, that $G_1 \oplus G_2 = G_2 \oplus G_1$. In general, the order in which the subgroups G_i are written in the symbol for their direct sum is immaterial.

As a simple illustration of a general property, suppose that G_1 and G_2 are subgroups of G such that $G = G_1 \oplus G_2$. Now if $G_1 = H_1 \oplus H_2$ and $G_2 = K_1 \oplus K_2$, where H_1 and H_2 are subgroups of G_1, and K_1 and K_2 are subgroups of G_2, then all of H_1, H_2, K_1, K_2 are subgroups of G and

$$G = H_1 \oplus H_2 \oplus K_1 \oplus K_2.$$

It will be clear that a similar result holds for any number of summands. (See Exercise 3 on p. 192.) This fact will be useful later on.

REMARK. To avoid any possible confusion, let us point out that if $G_1, G_2, \cdots, G_r$ are *any* additively written abelian groups (not necessarily subgroups of a given group), according to Section 7.1 the direct sum of these groups would consist of all ordered r-tuples

$$(a_1, a_2, \cdots, a_r),$$

with $a_i \in G_i$ for $i = 1, 2, \cdots, r$, and with addition defined as follows:

$$(a_1, a_2, \cdots, a_r) + (b_1, b_2, \cdots, b_r) = (a_1 + b_1, a_2 + b_2, \cdots, a_r + b_r).$$

Now if the G_i are subgroups of a group G and their sum is direct as defined in 8.1, it is not difficult to prove that the mapping

$$(a_1, a_2, \cdots, a_r) \to a_1 + a_2 + \cdots + a_r$$

is an isomorphism of the direct sum as defined in Section 7.1 onto the direct sum as defined in this section. Prove it! This fact justifies the use of the term *direct sum* in two different situations. Sometimes, the direct sum as defined in 8.1 is called an *internal* direct sum (since all groups G_i are subgroups of a given group G and therefore their direct sum is a subgroup of G), and the direct sum of Section 7.1 is called an *external* direct sum (since the G_i are arbitrary groups, not given as subgroups of some given group).

Now let G be an abelian group of order $n = p_1^{e_1}p_2^{e_2} \cdots p_k^{e_k}$, where the p's are distinct primes, $k \geq 1$, and each $e_i \geq 1$. Thus p_1, $p_2, \cdots, p_k$ are the distinct prime divisors of n. Let $G(p_i)$ be the set of all elements of G having order a power of p_i. The order of the identity 0 of G is $1 = p_i^0$ and hence $0 \in G(p_i)$. Actually, $G(p_i)$ is a subgroup of G since one can see as follows that $G(p_i)$ is closed under addition. If $a, b \in G(p_i)$, suppose that a has order p_i^m and b has order p_i^n. If t is the larger of n and m, then $p_i^t(a + b) = 0$, and the order of $a + b$ is a divisor of p_i^t; hence is a power of p_i. It follows that $a + b \in G(p_i)$, and by Theorem 7.4 we see that $G(p_i)$ is a subgroup of G. Our next goal is to prove the following theorem.

8.2 Theorem. *Let G be an abelian group of order n, and let p_1, $p_2, \cdots, p_k$ be the distinct prime divisors of n. If $G(p_i)$ is the subgroup of G consisting of all elements having order a power of p_i, then*

8.3
$$G = G(p_1) \oplus G(p_2) \oplus \cdots \oplus G(p_k).$$

Much later in this chapter we shall prove that no one of the subgroups $G(p_i)$ consists of the zero alone. In fact, if $p_i^{e_i}$ is the highest power of p_i which divides n, we shall show that $G(p_i)$ has order $p_i^{e_i}$.

Before proving Theorem 8.2, let us introduce some lemmas that will be helpful in carrying out the proof. It should be noted that in these lemmas n is temporarily being used to denote the order of an element, not the order of the group.

8.4 Lemma. *Suppose that the element a of an abelian group G has order n. If m is an integer such that $(m, n) = 1$, then $ma = 0$ implies that $a = 0$.*

PROOF. Since $(m, n) = 1$, there exist integers x and y such that $1 = xm + yn$. Hence $a = xma + yna$. We are assuming that $ma = 0$, and $na = 0$ since a has order n. It follows that $a = 0$, as we wished to show.

8.5 Lemma. *If the element a of the abelian group G has order $n = kl$ with $(k, l) = 1$, then there exist elements b and c of G such that $a = b + c$, with b and c having respective orders k and l.*

PROOF. Since $(k, l) = 1$, there exist integers s and t such that $1 = sk + tl$. Thus we have $a = ska + tla$. Let us show that ska has order l. Clearly, $lska = sna = 0$. Moreover, if $z \in \mathbf{Z}$ such that

$zska = 0$, then $n|zsk$. But $n = kl$, so we conclude that $kl|zsk$ or $l|zs$. Now the equation $1 = sk + tl$ implies that $(s, l) = 1$ and therefore $l|z$. Accordingly, we conclude that ska has order l. Similarly, tla has order k and if we set $b = tla$ and $c = ska$, we have $a = b + c$, b of order k and c of order l. This completes the proof.

We leave as an exercise the proof by induction of the following generalization of the preceding lemma.

8.6 Lemma. *If the element a of the abelian group G has order $n = n_1 n_2 \cdots n_k$, where $(n_i\ n_j) = 1$ for $i \neq j$, then a can be expressed in the form*

$$a = b_1 + b_2 + \cdots + b_k,$$

where b_i has order n_i $(i = 1, 2, \cdots, k)$.

PROOF OF 8.2. Let us now return to the proof of Theorem 8.2 and show first that the sum $G(p_i) + G(p_2) + \cdots + G(p_k)$ is a direct sum. To this end, suppose that

8.7 $$a_1 + a_2 + \cdots + a_k = 0, \qquad a_i \in G(p_i).$$

By definition of direct sum, we need to prove that each $a_i = 0$. For convenience of notation, let us concentrate on proving that $a_1 = 0$. Each a_i has order a power of p_i, so let us assume that a_i has order $p_i^{t_i}$ $(i = 1, 2, \cdots, k)$. From Equation 8.7, it now follows that

$$p_2^{t_2} p_3^{t_3} \cdots p_k^{t_k} a_1 = 0.$$

Since this coefficient of a_1 is relatively prime to the order of a_1, it follows from Lemma 8.4 that $a_1 = 0$. Similarly, each $a_i = 0$, and this proves that the sum is direct.

Clearly, $G(p_1) \oplus G(p_2) \oplus \cdots \oplus G(p_k) \subseteq G$, so we only need to obtain inclusion the other way. By Corollary 7.36, every element a of G has order a divisor of the order n of G, and therefore the order of a has no prime divisors except for some or all of the p_i $(i = 1, 2, \cdots, k)$. For convenience of notation only, suppose that the order of a contains only the prime factors $p_1, p_2, \cdots, p_u$, $u \leq k$. By Lemma 8.6, a is expressible as a sum of elements of $G(p_i)$, $i = 1, 2, \cdots, u$. In particular, every element of G is a sum of elements of some or all of the $G(p_i)$, $i = 1, 2, \cdots, k$. We therefore conclude that $G \subseteq G(p_i) \oplus G(p_2) \oplus \cdots \oplus G(p_k)$, and this completes the proof of the theorem.

8.2 CYCLIC SUBGROUPS AND BASES

We shall continue to let G be a finite abelian group. If $a \in G$, let us denote by (a) the cyclic subgroup generated by a. If a has order n, then

$$(a) = \{0, a, 2a, \cdots, (n-1)a\}.$$

8.8 Definition. If $a_1, a_2, \cdots, a_k$ are nonzero elements of G such that the sum $(a_1) + (a_2) + \cdots + (a_k)$ is direct, we say that the elements $a_1, a_2, \cdots, a_k$ are *independent* or form an *independent set.*

Suppose that a_i has order n_i $(i = 1, 2, \cdots, k)$. Then, by definition of direct sum, the a_i $(i = 1, 2, \cdots, k)$ are independent if for integers z_i,

$$z_1a_1 + z_2a_2 + \cdots + z_ka_k = 0$$

if and only if $z_ia_i = 0$, that is, if and only if $n_i | z_i$ $(i = 1, 2, \cdots, k)$.

Observe that a single element a of G is independent if and only if $a \neq 0$. Clearly, any nonempty subset of an independent set is also independent.

8.9 Definition. The set $\{a_1, a_2, \cdots, a_r\}$ forms a *basis* of the abelian group G if and only if the elements of this set are independent and

$$G = (a_1) \oplus (a_2) \oplus \cdots \oplus (a_r).$$

Otherwise expressed, the group G has a basis if and only if it is expressible as a direct sum of a finite number of cyclic subgroups.

We may remark that if $G = H_1 \oplus H_2$, where H_1 is a subgroup of G having basis $\{b_1, b_2, \cdots, b_s\}$ and H_2 is a subgroup of G having basis $\{c_1, c_2, \cdots, c_t\}$, then G has a basis $\{b_1, b_2, \cdots, b_s, c_1, c_2, \cdots, c_t\}$. (See Exercise 6 on p. 192.)

One of the principal theorems which we shall eventually prove is the following.

8.10 Theorem. *Every finite abelian group has a basis, each element of which has order a power of a prime.*

In view of our definitions, an equivalent formulation of this theorem would be the assertion that every finite abelian group can be expressed as the direct sum of cyclic subgroups, each of which has order a power of a prime.

By a generalization of the remark made above, the result of Theorem 8.2 shows that Theorem 8.10 will be true in general when we have established it for each of the groups $G(p_i)$. In the next section we study in some detail a class of groups which will include those of the form $G(p_i)$ as defined in Theorem 8.2.

EXERCISES

1. Prove the equivalence of conditions (i) and (ii) in connection with Definition 8.1.
2. Suppose that G_i $(i = 1, 2, \cdots, r)$ are subgroups of the abelian group G such that the sum $G_1 + G_2 + \cdots + G_r$ is direct. If, for each i, H_i is a subgroup of G_i, prove that the sum $H_1 + H_2 + \cdots + H_r$ is direct.
3. Suppose that $G = G_1 \oplus G_2$, where $G_1 = H_1 \oplus H_2 \oplus H_3$ and $G_2 = K_1 \oplus K_2$. Prove that $G = H_1 \oplus H_2 \oplus H_3 \oplus K_1 \oplus K_2$. Choose an appropriate notation and generalize to an arbitrary finite number of summands.
4. Let G be the additive group of the ring $\mathbf{Z}_{24}$. In the notation of Theorem 8.2, determine the elements of $G(2)$ and of $G(3)$. Verify that these are subgroups of G and that $G = G(2) \oplus G(3)$, thus directly verifying Theorem 8.2 for this particular group.
5. Give an example to show that in a finite nonabelian group the elements which have order a power of some fixed prime need not be a subgroup.
6. Suppose that G_1 and G_2 are subgroups of the abelian group G such that $G = G_1 \oplus G_2$. If $\{a_1, a_2, \cdots, a_r\}$ is a basis of G_1 and $\{b_1, b_2, \cdots, b_s\}$ is a basis of G_2, prove that $\{a_1, a_2, \cdots, a_r, b_1, b_2, \cdots, b_s\}$ is a basis of G. Generalize to direct sums of an arbitrary finite number of subgroups.
7. Illustrate Theorem 8.10 by verifying that for the group G which is the additive group of the ring $\mathbf{Z}_{24}$, a basis of the required kind is $\{3, 8\}$.
8. Prove Lemma 8.6.
9. Prove that a cyclic group of order p^k, where p is a prime and $k \geq 1$, cannot be expressed as a direct sum of two nonzero subgroups. [Hint: Consider the maximal order that an element can have.]

10. If (a) is a cyclic group of order kl with $(k, l) = 1$, prove that there exist elements b and c of (a) of respective orders k and l, such that $(a) = (b) \oplus (c)$.

11. If b and c are elements of an abelian group G with orders k and l respectively, and if $(k, l) = 1$, prove that the sum $(b) + (c)$ is direct and that $(b) \oplus (c)$ is a cyclic subgroup of G of order kl.

8.3 FINITE ABELIAN p-GROUPS

Let us begin with the following definition.

8.11 Definition. Let p be a fixed prime. A group is said to be a *p-group* if the order of each of its elements is a power of p.

We may observe that the identity element (the zero) has order p^0. Every other element of a p-group has order p^m for some positive integer m. Thus a nonzero element a of a p-group has order p^m if and only if $p^m a = 0$, $p^{m-1}a \neq 0$. Moreover, if a has order p^m and the order of an element b of the p-group is less than or equal to the order of a, then $p^m b = 0$.

The main goal of this section is to prove the following special case of Theorem 8.10.

8.12 Lemma. *A finite abelian p-group has a basis.*

Throughout this section, let p be a fixed prime and G a finite abelian p-group. As a first step in the proof of Lemma 8.12, we collect a few useful facts for easy reference.

Let H be a subgroup of G and suppose that a is an element of G of order p^m. Since $p^m a = 0 \in H$, there exists a smallest positive integer z (necessarily less than or equal to p^m) such that $za \in H$. Throughout this section it will be convenient to have a distinctive name for this positive integer z. We shall call it the *degree of a relative to H*. Thus, the order of a is the degree of a relative to the zero subgroup.

(A) (*i*) *If z is the degree of a relative to H and $n \in \mathbf{Z}$, then $na \in H$ if and only if $z | n$. In particular, if z is the degree of a relative to H and $za = 0$, then z is the order of a.*
(*ii*) *If a has order p^m, and z is the degree of a relative to H, then $z | p^m$ and therefore the degree of each element of G relative to any subgroup of G is a power of p.*

PROOF. To prove (i), let us use the Division Algorithm to write $n = qz + r$, where $0 \leq r < z$. Thus $na = q(za) + ra$. Now $za \in H$, so if $na \in H$, it follows that $ra \in H$. Since z is the smallest positive integer such that $za \in H$, we conclude that $r = 0$ and therefore $n = qz$. Conversely, if $z|n$, it is trivial that $na \in H$. The last statement of (i) follows from the observation that the given conditions imply that z and the order of a divide each other.

Part (ii) follows from (i) by observing that $p^m a = 0 \in H$, and therefore $z|p^m$ and z must therefore be a power of p.

(**B**) *Suppose that H is a subgroup of G and that $a \notin H$. If the order p^m of a is equal to the degree of a relative to H, then the sum $H + (a)$ is direct.*

PROOF. To see the truth of this statement, suppose that $h + xa = 0$, where $h \in H$ and $x \in \mathbf{Z}$, and let us prove that $h = 0$ and that $xa = 0$. Now $xa \in H$ and since a has degree p^m relative to H, part (i) of (A) shows that $p^m|x$. But this implies that $xa = 0$ since p^m is also the order of a. The equation $h + xa = 0$ then shows that $h = 0$. The sum $H + (a)$ is therefore a direct sum, as we wished to show.

The proof of Lemma 8.12 is carried out by the process of induction. We illustrate the approach by a fairly detailed account of the first two steps in this procedure, and then pass on to the general situation.

Let a_1 be an element of G of maximum order, say p^{m_1}. If it happens that $G = (a_1)$, we have found a basis $\{a_1\}$ of G consisting of the single element a_1. Suppose, then, that $G \neq (a_1)$. Since a_1 has maximum order among the elements of G, we see that $p^{m_1}c = 0$ for *every* element c of G.

Since $G \neq (a_1)$, we proceed to seek an element a_2 of G such that a_1, a_2 are independent and therefore $(a_1) \oplus (a_2) \subseteq G$.

Let b be an element of G of maximum degree, say p^{m_2}, relative to (a_1). Since $p^{m_1}b = 0 \in (a_1)$, it follows that $p^{m_2} \leq p^{m_1}$, that is, $m_2 \leq m_1$. Since b has maximum degree relative to (a_1), we see that $p^{m_2}c \in (a_1)$ for every element c of G. Thus if $c \in G$, there exists $y \in Z$ such that

8.13 $$p^{m_2}c = ya_1.$$

We proceed to show that $p^{m_2}|y$. Multiplying the preceding equation by $p^{m_1-m_2}$, we find that $p^{m_1}c = p^{m_1-m_2}ya_1$. But $p^{m_1}c = 0$, and since a_1 has order p^{m_1}, we conclude that $p^{m_1}|p^{m_1-m_2}y$, that is, that $p^{m_2}|y$. Thus $y = up^{m_2}$, where $u \in \mathbf{Z}$. Now applying what we have just proved to the special case in which c is an element b of maximum degree relative to (a_1), we find from 8.13 that there exists $u \in \mathbf{Z}$ such that

8.14 $$p^{m_2}b = p^{m_2}ua_1.$$

We now set

8.15 $$a_2 = b - ua_1,$$

and observe that if $z \in \mathbf{Z}$, then $za_2 \in (a_1)$ if and only if $zb \in (a_1)$. Thus the degree of a_2 relative to (a_1) is p^{m_2}. Now 8.14 and 8.15 imply that $p^{m_2}a_2 = 0$, and it follows from (A)(i) that the order of a_2 is p^{m_2}. We now know from (B) that the sum $(a_1) + (a_2)$ is a direct sum, and we have $G \supseteq (a_1) \oplus (a_2)$. If $G = (a_1) \oplus (a_2)$, we have exhibited a basis $\{a_1, a_2\}$. Otherwise, we can continue this process. We have just completed the case $k = 2$ of the following induction procedure.

Assume that we have found elements $a_1, a_2, \cdots, a_k$ of G of respective orders $p^{m_1}, p^{m_2}, \cdots, p^{m_k}$ such that all of the following are true:

(i) $m_1 \geq m_2 \geq \cdots \geq m_k$.
(ii) If $c \in G$, there exist integers $y_1, \cdots, y_{k-1}$ such that $p^{m_k}c = y_1a_1 + \cdots + y_{k-1}a_{k-1}$, and $p^{m_k}|y_i$ $(i = 1, 2, \cdots, k - 1)$.
(iii) $a_1, a_2, \cdots, a_k$ are independent.

For convenience, let us set $G_{k-1} = (a_1) \oplus (a_2) \oplus \cdots \oplus (a_{k-1})$ and $G_k = (a_1) \oplus (a_2) \oplus \cdots \oplus (a_k)$. If $G \neq G_k$, we propose to find another element a_{k+1} of G such that all of the above three properties are true with k replaced by $k + 1$.

We may point out that (ii) above states not only that if $c \in G$, then $p^{m_k}c \in G_{k-1}$, but even gives some additional information in that the integers y_i are all divisible by p^{m_k}.

Let b be an element of G of maximum degree, say $p^{m_{k+1}}$, relative to G_k. Thus, if $c \in G$, then $p^{m_{k+1}}c \in G_k$. Observe that since $p^{m_k}b \in G_{k-1} \subseteq G_k$, $p^{m_{k+1}} \leq p^{m_k}$ and $m_k \geq m_{k+1}$, so (i) holds for $k + 1$.

If $c \in G$, since $p^{m_{k+1}}c \in G_k$, there exist integers $z_1, \cdots, z_k$ such that

8.16 $$p^{m_{k+1}}c = z_1a_1 + z_2a_2 + \cdots + z_ka_k.$$

We propose to show that $p^{m_{k+1}} | z_i$ for $i = 1, 2, \cdots, k$. By (ii) of our induction hypothesis, there exist integers $y_1, \cdots, y_{k-1}$ such that

8.17 $$p^{m_k}c = y_1a_1 + \cdots + y_{k-1}a_{k-1}, \quad p^{m_k}|y_i \ (i = 1, 2, \cdots, k).$$

If we multiply 8.16 by $p^{m_k - m_{k-1}}$, we obtain $p^{m_k}c = p^{m_k - m_{k+1}}z_1a_1 + \cdots + p^{m_k - m_{k+1}}z_ka_k$. By equating the right sides of the two preceding equations, we obtain

$$(p^{m_k - m_{k+1}}z_1 - y_1)a_1 + \cdots + (p^{m_k - m_{k+1}}z_{k-1} - y_{k-1})a_{k-1} + p^{m_k - m_{k+1}}z_ka_k = 0.$$

Since, by assumption, $a_1, a_2, \cdots, a_k$ are independent, for each i the coefficient of a_i in this equation must be divisible by the order of a_i. In particular, $p^{m_k}|p^{m_k - m_{k+1}}z_k$ and this implies that $p^{m_{k+1}}|z_k$. Now for $1 \leq i < k$, we have $p^{m_i}|(p^{m_k - m_{k+1}}z_i - y_i)$. But $i < k$ and so $m_i \geq m_k$, and thus

$$p^{m_k}|(p^{m_k - m_{k+1}}z_i - y_i).$$

But as indicated in 8.17, we know that $p^{m_k}|y_i$. It follows that $p^{m_k}|p^{m_k - m_{k+1}}z_i$ and hence that $p^{m_{k+1}}|z_i$. Since this is true for $1 \leq i < k$, and we have already shown that $p^{m_{k+1}}|z_k$, we conclude that $p^{m_{k+1}}$ divides every z_i in 8.16. This establishes part (ii) of our induction statement for the case in which k is replaced by $k + 1$.

Now let us apply what we have just proved to the special case in which the element c in 8.16 is chosen to be a particular element b of maximal degree $p^{m_{k+1}}$ relative to G_k. Thus, since $p^{m_{k+1}}|z_i$, there exist integers $u_1, \cdots, u_k$ such that

8.18 $$p^{m_{k+1}}b = p^{m_{k+1}}u_1a_1 + p^{m_{k+1}}u_2a_2 + \cdots + p^{m_{k+1}}u_ka_k.$$

We next define

8.19 $$a_{k+1} = b - u_1a_1 - \cdots - u_ka_k,$$

and observe from this equation that a_{k+1} has the same degree relative to G_k as does b, namely, $p^{m_{k+1}}$. The last two equations show that $p^{m_{k+1}}a_{k+1} = 0$ and it follows from (A)(i) that the order of a_{k+1} is $p^{m_{k+1}}$. By (B), the sum $G_k + (a_{k+1})$ is direct and it follows that $a_1, a_2, \cdots, a_{k+1}$ are independent. We have shown that if $a_1, a_2, \cdots, a_k$ satisfy the induction hypotheses (i), (ii), and (iii), and if $G \neq (a_1) \oplus \cdots \oplus (a_k)$, there exists an element a_{k+1} such that $a_1, \cdots, a_{k+1}$ satisfy (i), (ii), and (iii), with k replaced by $k + 1$. In particular, $G \supseteq (a_1) \oplus \cdots \oplus (a_{k+1})$. Since G is assumed to be a *finite* p-group, these steps must come to an end, and thus for some positive integer r, there exist elements $a_1, a_2, \cdots, a_r$ such that

8.20
$$G = (a_1) \oplus (a_2) \oplus \cdots \oplus (a_r).$$

This completes the proof of Lemma 8.12, which states that every finite abelian p-group has a basis.

In proving 8.20, we have obtained the basis elements $a_1, a_2, \cdots, a_r$ such that their orders are respectively $p^{m_1}, p^{m_2}, \cdots, p^{m_r}$ with $m_1 \geq m_2 \geq \cdots \geq m_r \geq 1$. It follows from 8.20 that the order of G is the product of the orders of the cyclic groups (a_i), namely p^t, where $t = m_1 + m_2 + \cdots + m_r$. In particular, this shows that *the order of a finite abelian p-group is a power of p.*

Before returning to the study of arbitrary finite abelian groups, let us discuss the question of the uniqueness of a basis for a p-group. Clearly, our construction of a basis indicates that a basis is *not* unique since, as a simple example, a_1 might have been chosen to be any element of maximal order. However, we shall prove the following result.

8.21 Theorem. *Any two bases of a finite abelian p-group have the same number of elements. Moreover, the orders of the elements of one basis coincide, in some arrangement, with the orders of the elements of any other basis.*

PROOF. In proving this theorem, we shall assume that the p-group G has a basis $\{a_1, a_2, \cdots, a_r\}$, with a_i having order p^{m_i}; and that G also has a basis $\{b_1, b_2, \cdots, b_s\}$, with b_i having order p^{n_i}. Moreover, we assume that the notation is chosen so that $m_1 \geq m_2 \geq \cdots \geq m_r \geq 1$ and $n_1 \geq n_2 \geq \cdots \geq n_s \geq 1$. We proceed to prove that $r = s$ and that $m_i = n_i$ $(i = 1, 2, \cdots, r)$.

It will be convenient to consider subgroups pG and G_p of G, defined as follows:

$$pG = \{px \mid x \in G\},$$
$$G_p = \{x \mid x \in G, px = 0\}.$$

Thus $pG = \{0\}$ if and only if $G_p = G$.

Making use of the basis $\{a_1, a_2, \cdots, a_r\}$ of G, we leave it to the reader to verify that

8.22 $$\{p^{m_1-1}a_1, p^{m_2-1}a_2, \cdots, p^{m_r-1}a_r\}$$

is a basis of the p-group G_p. Since each of these basis elements has order p and G_p is the direct sum of the cyclic groups generated by these elements, we conclude that G_p has order p^r. In exactly the same way, using the basis $\{b_1, b_2, \cdots, b_s\}$ of G, we see that G_p has order p^s. Hence $p^r = p^s$, and $r = s$. This completes the proof of the first statement of the theorem.

The proof of the second statement is by induction on the order of G, and we therefore assume as an induction hypothesis that the statement is true for all p-groups with order less than the order of G. We now make two cases, in the first of which we do not really need this induction hypothesis.

CASE 1. $pG = \{0\}$. In this case, every nonzero element (in particular, every basis element) of G has order p. Hence, $m_i = n_i = 1$ $(i = 1, 2, \cdots, r)$.

CASE 2. $pG \neq \{0\}$. In this case pG is a nonzero subgroup of G. Moreover, the order of pG is less than the order of G, since G necessarily has some elements of order p. Why? Using the notation in which the order of a_i is p^{m_i}, not all m_i can equal 1. Suppose that u is a positive integer so chosen that $m_1 \geq m_2 \geq \cdots \geq m_u > m_{u+1} = \cdots = m_r = 1$. It may now be verified that pG has a basis

8.23 $$\{pa_1, \cdots, pa_u\}.$$

In like manner, making use of the other given basis $\{b_1, \cdots, b_r\}$ of G, if v is the positive integer so chosen that $n_1 \geq n_2 \geq \cdots \geq n_v > n_{v+1} = \cdots = n_r = 1$, we see that pG has a basis

8.24 $$\{pb_1, \cdots, pb_v\}.$$

Thus the p-group pG has bases 8.23 and 8.24. By the first statement of the theorem, already proved, we conclude that $u = v$. Using the fact that the order of pa_i is p^{m_i-1} and the order of pb_i is p^{n_i-1}, the induction hypothesis as applied to the group pG shows that $m_i - 1 = n_i - 1$ for $i = 1, 2, \cdots, u$. Since all other m_i and n_i are equal to 1, we have that $m_i = n_i$ $(i = 1, 2, \cdots, r)$. This concludes the proof of the theorem.

8.4 THE PRINCIPAL THEOREMS FOR FINITE ABELIAN GROUPS

Henceforth we shall let G be an arbitrary finite abelian group. Let us assume that G has order n with distinct prime factors $p_1, p_2, \cdots, p_k$. Thus

8.25 $$n = p_1^{e_1}p_2^{e_2} \cdots p_k^{e_k},$$

where $e_i > 0$ for all i. If $G(p_i)$ denotes the subgroup of G consisting of all elements of order a power of p_i, we have proved in Theorem 8.2 that

8.26 $$G = G(p_1) \oplus \cdots \oplus G(p_k).$$

Now $G(p_i)$ is a p_i-group and, as indicated shortly before the statement of Theorem 8.21, its order is a power of p_i. Moreover, from Equation 8.26, the order n of G must be the product of the orders of the groups $G(p_i)$. In view of the unique factorization of n into a product of primes, we conclude from 8.25 that the order of $G(p_i)$ must be $p_i^{e_i}$ $(i = 1, 2, \cdots, k)$. We have therefore proved the first statement of the following lemma.

8.27 Lemma. *Let G be an abelian group of order n.*

(*i*) *If p is a prime divisor of n, let p^e be the highest power of p which divides n. Then the subgroup $G(p)$ of G which consists of all elements with order a power of p, has order p^e. In particular, $G(p) \neq \{0\}$.*
(*ii*) *If p is a prime divisor of n, then G contains an element of order p.*

PROOF. The proof of part (ii) follows at once from the observation that if a is a nonzero element of $G(p)$, then a has order p^t for some positive integer t. Hence $p^{t-1}a$ has order p.

The results of the preceding section show that each subgroup $G(p_i)$ occurring in 8.26 has a basis, say $\{a_{i1}, a_{i2}, \cdots, a_{ir_i}\}$, and clearly each element of this basis has order a power of p_i. Using this information, Equation 8.26 shows that G has a basis

8.28 $$\{a_{11}, a_{12}, \cdots, a_{1r_1}; a_{21}, a_{22}, \cdots, a_{2r_2}; \cdots; a_{k1}, a_{k2}, \cdots, a_{kr_k}\},$$

and each element of this basis has order a power of a prime. This result was stated as Theorem 8.10, one of our principal goals. We have therefore proved part (i) of the following fundamental theorem.

8.29 Fundamental Theorem on Finite Abelian Groups.

(*i*) *Every finite abelian group G has a basis, each element of which has order a power of a prime.*

(*ii*) *Suppose we have any two bases of a finite abelian group G, with each basis element having order a power of a prime. Then the two bases have the same number of elements and the orders of the elements of one basis are, in some arrangement, the same as the orders of the elements of the other basis.*

PROOF. To prove part (ii) of this theorem, suppose that one basis of G is given by 8.28. If G has order n, given by 8.25, the order of every element of G is a divisor of n and hence the only possible primes a power of which can occur as the order of any (basis) element are $p_1, p_2, \cdots, p_k$. Suppose that in a second basis of G, the elements $b_1, \cdots, b_t$ are those whose orders are a power of p_1. Then the elements of G whose elements are a power of p_1 are precisely the elements of the direct sum

8.30 $$(b_1) \oplus \cdots \oplus (b_t).$$

It follows that the direct sum 8.30 is equal to $G(p_1)$. Since from 8.28 we also have

8.31 $$G(p_1) = (a_{11}) \oplus \cdots \oplus (a_{1r_1}),$$

we may apply Theorem 8.21 to the p_1-group $G(p_1)$, and conclude that $t = r_1$, and that the orders of $b_1, \cdots, b_t$ coincide in some arrangement, with the orders of $a_{11}, \cdots, a_{1r_1}$. Thus the number of basis elements having order a power of p_1 is the same in the two bases, as are also the orders of the elements of the two bases. The same argument applies equally well to each prime p_i, and this completes the proof of the theorem.

Let us make the following definition.

8.32 Definition. Let G be a finite abelian group. The orders of the elements of a basis (repetitions being allowed), in which each basis element is required to have order a power of a prime, are called the *invariants* (or *elementary divisors*) of G.

Thus, for example, if we say that G has invariants 3, 2^2, 2, 2, it means that G is expressible as a direct sum of cyclic groups of these respective orders. Thus for this group G, we have

$$G \simeq C_3 \oplus C_{2^2} \oplus C_2 \oplus C_2,$$

where C_n represents a cyclic group of order n.

The concept of the invariants of an abelian group is important because of the following theorem.

8.33 Theorem. *Two finite abelian groups are isomorphic if and only if they have the same invariants.*

PROOF. One part of this result follows fairly easily from results obtained above. Suppose that $\theta: G \to G'$ is an isomorphism of the finite abelian group G onto the finite abelian group G'. If $\{a_1, a_2, \cdots, a_n\}$ is a basis of G, with each a_i having order a power of a prime, the orders of $a_1, a_2, \cdots, a_n$ are then the invariants of G. Now $\{a_1\theta, a_2\theta, \cdots, a_n\theta\}$ is a basis of G' (see Exercise 5 below), and under the isomorphism θ, a_i and $a_i\theta$ have the same order. Hence G' has the same invariants as G.

Conversely, suppose that G and G' have the same invariants. This means that

$$G \simeq D_1 \oplus D_2 \oplus \cdots \oplus D_n$$

and

$$G' \simeq E_1 \oplus E_2 \oplus \cdots \oplus E_n,$$

where D_i and E_i are cyclic groups of the same order (a power of a prime). Now, by Theorem 7.29, two cyclic groups of the same order are isomorphic. Let $\theta_i: D_i \to E_i$ be an isomorphism of D_i onto E_i. Then it may be verified that the mapping $\theta: G \to G'$ defined by

8.34 $$(d_1 + d_2 + \cdots + d_n)\theta = d_1\theta_1 + d_2\theta_2 + \cdots + d_n\theta_n,$$

where $d_i \in D_i$, is an isomorphism of G onto G'. (See Exercise 4 below.)

As a simple application of this theorem, let us determine all nonisomorphic abelian groups of order 24. Since the product of the invariants must be 24, we find the following possible systems of invariants: $3, 2^3$; $3, 2^2, 2$; $3, 2, 2, 2$. Thus there are three nonisomorphic abelian groups of order 24. If, as above, we let C_n denote a cyclic group of order n, these three nonisomorphic abelian groups of order 24 are respectively isomorphic to

$$C_3 \oplus C_{2^3}, \quad C_3 \oplus C_{2^2} \oplus C_2, \quad C_3 \oplus C_2 \oplus C_2 \oplus C_2.$$

EXERCISES

1. If an abelian group G has invariants 2^2, 5, 5; verify that there exist elements G of order 20 and none of higher order. Determine the number of elements of order 20.
2. In the notation used in the proof of Theorem 8.21, prove that the set 8.22 is a basis of G_p.
3. In the notation of the same proof, prove that the set 8.23 is a basis of pG.
4. Verify that the mapping θ defined in 8.34 is an isomorphism of G onto G'.
5. Prove: If $\{a_1, a_2, \cdots, a_n\}$ is a basis of an abelian group G and if $\theta: G \to G'$ is an isomorphism of G onto G', then $\{a_1\theta, a_2\theta, \cdots a_n\theta\}$ is a basis of G', and G and G' have the same invariants.
6. If $p_1, p_2, \cdots, p_k$ are distinct primes, show that any two abelian groups of order $p_1p_2 \cdots p_k$ are isomorphic (and therefore isomorphic to the cyclic group of this order).
7. Suppose that an abelian group G has order $n = p_1^{e_1}p_2^{e_2} \cdots p_k^{e_k}$, where the p's are distinct primes and each $e_i \geq 1$. If among the invariants of G the highest powers of these primes which occur are $p_1^{t_1}$, $p_2^{t_2}$, $\cdots$, $p_k^{t_k}$, prove that there exists an element of G of order $p_1^{t_1}p_2^{t_2} \cdots p_k^{t_k}$ and no element of higher order. [See Exercise 14 at the end of Section 7.4. Observe also that Exercise 1 above involves a verification of a special case of this result.]
8. Verify that there are exactly four nonisomorphic abelian groups of order 100. For each of these groups, determine the maximal order of an element.
9. Show that if a cyclic group G has order p^m (p a prime) and if $t \in \mathbf{Z}$ such that $0 \leq t \leq m$, then G has a subgroup of order p^t.
10. Prove that if p and t are as in the preceding exercise, any abelian group G of order p^m has a subgroup of order p^t. [Hint: Consider the invariants of G, the result of the preceding exercise, and Exercise 2 of the preceding set.]
11. Use Theorem 8.2, Lemma 8.27, and the results of the two preceding exercises to prove the following general result: If an abelian group G has order n and $k|n$, then G has a subgroup of order k.

NOTES AND REFERENCES

This chapter is essentially the same as Chapter 8 of McCoy [8]. A variety of proofs of the fundamental theorem on finite abelian groups (8.29) will be found in the books on group theory which are listed in the Bibliography, and also in some of the books on abstract algebra in general. Important generalizations to what are called "finitely generated abelian groups" will be found, e.g., in Lederman [16], Baumslag and Chandler [4], and Rotman [17].

Chapter 9

Polynomials

In elementary algebra an important role is played by polynomials in a symbol "x" with coefficients that are real or complex numbers. In the next section we shall introduce polynomials with coefficients in a commutative ring S with unity, and show that under suitable definitions of addition and multiplication the set of all such polynomials is a ring. Actually, we could just as well start by letting S be an entirely arbitrary ring, but in most of the chapter it is essential that it be commutative and so we simplify matters by making this assumption from the beginning. The restriction that S have a unity is not very important but it does serve to simplify the notation somewhat.

The purpose of this chapter is to introduce polynomials with coefficients in a commutative ring S with unity, and to establish a number of properties of such polynomials. We shall frequently find it necessary or desirable to make additional restrictions on the ring S. In particular, we shall sometimes require that it be a field or a specified one of the fields that have already been studied in detail in previous chapters.

It will be found that a ring of polynomials with coefficients in a *field* has a considerable number of properties in common with the ring **Z** of integers. Accordingly, several of the sections of this chapter will closely parallel corresponding material of Chapter 4.

9.1 POLYNOMIAL RINGS

Let S be a commutative ring with unity. Heretofore we have used letters to denote sets or elements of sets, but we now use the letter x in a different way. It is not an element of S, but is just a symbol which we shall use in an entirely formal way. It is customary to call such a symbol an *indeterminate*. It is our purpose in this section to construct a ring which contains S and also has x as an element. This goal will motivate the definitions which we proceed to give.

Let x be an indeterminate and let us consider expressions of the form

9.1
$$a_0x^0 + a_1x^1 + a_2x^2 + \cdots + a_nx^n,$$

where n is some nonnegative integer and $a_i \in S$ ($i = 0, 1, \cdots, n$). Such an expression is called "a *polynomial* in x with coefficients in S" or simply "a polynomial in x over S." If i is an integer such that $0 \leq i \leq n$, we say that a_i is the *coefficient* of x^i in the polynomial 9.1; also we say that a_ix^i is a *term* of the polynomial 9.1 with coefficient a_i.

At this stage we are to think of 9.1 as a purely formal expression. That is, the + signs are not to be considered as representing addition in a ring, and neither is x^i to be considered as a product $x \cdot x \cdots x$ with i factors. Later on, after we have proved the existence of a ring which contains S as well as x, we shall see that in this larger ring we can make these familiar interpretations and thus justify the notation we are using. At the present time, we could logically use some such symbol as $(a_0, a_1, a_2, \cdots, a_n)$ to designate the polynomial 9.1, but the definitions of addition and multiplication of polynomials to be given below will seem more natural with the familiar notation used in 9.1.

For the moment, let S be the ring $\mathbf{Z}$ of integers. Then the following are examples of polynomials in x over $\mathbf{Z}$:

(i) $2x^0 + (-3)x^1 + 4x^2$,
(ii) $3x^0$,
(iii) $0x^0 + 0x^1 + 4x^2$,
(iv) $0x^0 + 2x^1 + (-1)x^2 + 0x^3$.

In order to avoid writing so many terms with zero coefficients, we could agree in the third of these examples to write merely $4x^2$ with the understanding that x^0 and x^1 are assumed to have zero coefficients. Also, it would certainly agree with usual practice if we omitted the terms with zero coefficients in the fourth example and wrote $2x^1 + (-1)x^2$ to designate this polynomial. These simplifications will be possible under general agreements which we now make.

Let us designate the polynomial 9.1 over S by the symbol $f(x)$, and let $g(x)$ be the following polynomial over S:

9.2 $$b_0x^0 + b_1x^1 + \cdots + b_mx^m,$$

where $m \geq 0$ and $b_i \in S$ $(i = 0, 1, \cdots, m)$. By the *equality* of $f(x)$ and $g(x)$, written in the usual way as $f(x) = g(x)$, we shall mean that the expressions 9.1 and 9.2 are identical except for terms with zero coefficients. We therefore consider a polynomial as being unchanged by the insertion, or omission, of any number of terms with zero coefficients. In particular, with reference to the above examples, we may write

$$0x^0 + 0x^1 + 4x^2 = 4x^2$$

and

$$0x^0 + 2x^1 + (-1)x^2 + 0x^3 = 2x^1 + (-1)x^2.$$

Also, if we wish, we could write

$$3x^0 = 3x^0 + 0x^1 + 0x^2 + 0x^3,$$

and so on.

With this understanding about zero coefficients, if $f(x)$ is a polynomial over S and i is an *arbitrary* nonnegative integer, we may speak of the coefficient of x^i in $f(x)$. For example, in the polynomial $1x^0 + 2x^1 + 3x^2$ over $\mathbf{Z}$, the coefficient of x^{10} is zero. This language often helps to simplify statements about polynomials. As an illustration, we may state again our definition of equality of two polynomials as follows. If $f(x)$ and $g(x)$ are polynomials over S, by $f(x) = g(x)$ we mean that for *every* nonnegative integer i, the coefficients of x^i in $f(x)$ and in $g(x)$ are equal elements of S.

Using the familiar sigma notation for sums, the polynomial 9.1 can be formally written as follows:

$$\sum_{i=0}^{n} a_ix^i.$$

Moreover, in view of our agreement about zero coefficients, we can write an arbitrary polynomial in x over S in the form

$$\sum_{i=0} a_ix^i,$$

with the tacit understanding that all coefficients are zero from some point on, so that this sum may be considered to be a finite sum with an unspecified number of terms.

Now let $S[x]$ denote the set of all polynomials in the indeterminate x over S. We proceed to define operations of addition and multiplication on the set $S[x]$. Of course, these definitions are suggested by the way that one adds and multiplies polynomials in elementary algebra. Let

9.3
$$f(x) = \sum_{i=0} a_i x^i$$

and

9.4
$$g(x) = \sum_{i=0} b_i x^i$$

be elements of $S[x]$. We define addition as follows:

9.5
$$f(x) + g(x) = \sum_{i=0} (a_i + b_i)x^i;$$

that is, for every nonnegative integer i, the coefficient of x^i in $f(x) + g(x)$ is the sum of the coefficients of x^i in $f(x)$ and in $g(x)$. Multiplication in $S[x]$ is defined as follows:

9.6
$$f(x)g(x) = \sum_{i=0} \left(\sum_{k=0}^{i} a_k b_{i-k} \right) x^i.$$

Another way of stating this definition of the product of $f(x)$ and $g(x)$ is to say that for each nonnegative integer i, the coefficient of x^i in the product is the sum (in the ring S) of all products of the form $a_r b_s$, where r and s are nonnegative integers such that $r + s = i$. The first few terms in the product given in 9.6 are as follows:

$$(a_0b_0)x^0 + (a_0b_1 + a_1b_0)x^1 + (a_0b_2 + a_1b_1 + a_2b_0)x^2 + \cdots.$$

We are now ready to state the following theorem.

9.7 Theorem. *Let $S[x]$ be the set of all polynomials in the indeterminate x over the commutative ring S with unity. If operations of addition and multiplication are defined on $S[x]$ by 9.5 and 9.6, respectively, then*

(*i*) *$S[x]$ is a commutative ring with unity,*
(*ii*) *$S[x]$ contains a subring isomorphic to S,*
(*iii*) *$S[x]$ is an integral domain if and only if S is an integral domain.*

PROOF. The commutative and associative laws for addition in $S[x]$ follow from 9.5 since these laws hold in the ring S. Moreover, the polynomial $0x^0$ (which is equal to the polynomial with *all* coefficients zero) is the zero of $S[x]$ since, by 9.5, for each polynomial $f(x)$ we have

$$f(x) + 0x^0 = f(x).$$

Moreover, our definition of addition also shows that

$$\sum_{i=0} a_i x^i + \sum_{i=0} (-a_i)x^i = \sum_{i=0} [a_i + (-a_i)]x^i = 0x^0,$$

and each element of $S[x]$ has an additive inverse in $S[x]$.

To establish that multiplication is commutative, we observe that if $f(x)$ and $g(x)$ are given by 9.3 and 9.4, respectively, then the coefficient of x^i in $g(x)f(x)$ is

$$b_0 a_i + b_1 a_{i-1} + \cdots + b_i a_0,$$

and since S is assumed to be commutative, this is equal to the coefficient

$$a_0 b_i + a_1 b_{i-1} + \cdots + a_i b_0$$

of x^i in $f(x)g(x)$. Inasmuch as this statement is true for every nonnegative integer i, it follows that $f(x)g(x) = g(x)f(x)$, and hence that multiplication in $S[x]$ is commutative.

If $f(x)$ and $g(x)$ are given by 9.3 and 9.4, respectively, and

9.8 $$h(x) = c_0 x^0 + c_1 x^1 + \cdots + c_p x^p$$

is also an element of $S[x]$, the coefficient of x^i in the product $(f(x)g(x))h(x)$ is found to be the sum of all products of the form $(a_r b_s)c_t$, where r, s, and t are nonnegative integers such that $r + s + t = i$. Similarly, the coefficient of x^i in the product $f(x)(g(x)h(x))$ is the sum of all products of the form $a_r(b_s c_t)$, with the same restriction on r, s, and t. However, since $(a_r b_s)c_t = a_r(b_s c_t)$ by the associative law of multiplication in S, it follows that

$$(f(x)g(x))h(x) = f(x)(g(x)h(x)),$$

that is, that multiplication is associative in $S[x]$.

We leave as exercises the proof of the distributive laws, and that if 1 is the unity of S, then $1x^0$ is the unity of $S[x]$. It follows then that $S[x]$ is a commutative ring with unity.

To establish part (ii) of the theorem, let S' denote the set of elements of $S[x]$ of the form ax^0, $a \in S$. It is easy to verify that S' is a subring of $S[x]$. Now the mapping $\theta: S' \to S$ defined by $(ax^0)\theta = a$, $a \in S$, is a one-one mapping of S' onto S. Moreover,

$$(ax^0 + bx^0)\theta = [(a + b)x^0]\theta = a + b = (ax^0)\theta + (bx^0)\theta$$

and

$$[(ax^0)(bx^0)]\theta = [(ab)x^0]\theta = ab = [(ax^0)\theta][(bx^0)\theta],$$

and it follows that θ is an isomorphism of S' onto S. Part (ii) of the theorem is therefore established.

Before proceeding to the proof of part (iii) of the theorem, let us introduce some simplifications of our notation as follows. We shall henceforth identify S' with S, and therefore write simply a in place of ax^0; that is, we shall omit x^0 in writing polynomials. In particular, the zero polynomial will then be designated by the familiar symbol 0. We shall also write x in place of x^1, x^i in place of $1x^i$, and $-ax^i$ in place of $(-a)x^i$. We may now observe that x is itself an element of the ring $S[x]$. If $a \in S$, then also $a \in S[x]$, and ax^i can be interpreted as the product (in the ring $S[x]$) of a times x to the power i. Also, since each individual term of a polynomial 9.1 is itself equal to a polynomial, the $+$ signs occurring in 9.1 can be correctly interpreted as addition in the ring $S[x]$. In other words, we have finally justified the use of the notation appearing in 9.1. Of course, addition is commutative in $S[x]$ and we can write the polynomial 9.1 with the terms in any order. For example, we could just as well write the polynomial 9.1 in the form

$$a_nx^n + a_{n-1}x^{n-1} + \cdots + a_1x + a_0.$$

In this case it is customary to say that it is written in *decreasing* powers of x. As given in 9.1, it is written in *increasing* powers of x.

The following familiar concepts are of such great importance that we give a formal definition.

9.9 Definition. Let $f(x)$ be a nonzero element of the ring $S[x]$. If n is the largest nonnegative integer such that x^n has a nonzero coefficient in $f(x)$, we say that $f(x)$ has *degree n*. If $f(x)$ has degree n, the nonzero coefficient of x^n is sometimes called the *leading coefficient* of $f(x)$. The zero polynomial has no degree and therefore also no leading coefficient. The coefficient of x^0 in a polynomial—that is, as now written, the term that does not involve x—is sometimes referred to as the *constant term* of the polynomial.

It will be observed that the nonzero elements of S, considered as elements of $S[x]$, are just the polynomials of degree zero. The degree of a polynomial $f(x)$ may be conveniently designated by $\deg f(x)$.

If S is the ring $\mathbf{Z}$ of integers, the polynomials $2 + 3x - x^2$, $4x$, 3, and $x^4 - 2x$ have respective degrees 2, 1, 0, and 4 and respective leading coefficients -1, 4, 3, and 1. The constant terms are, respectively, 2, 0, 3, and 0.

The third part of Theorem 9.7 will follow immediately from the following lemma.

9.10 Lemma. *Let S be an integral domain and let $f(x)$ and $g(x)$ be nonzero elements of $S[x]$. Then*

9.11 $$\deg\,(f(x)g(x)) = \deg f(x) + \deg g(x)\,.$$

PROOF. Since $f(x)$ and $g(x)$ are not zero, they have degrees, and let us suppose that $\deg f(x) = n$ and $\deg g(x) = m$. Then $f(x)$ can be written in the form 9.3 with $a_n \neq 0$ and $g(x)$ in the form 9.4 with $b_m \neq 0$. It now follows by the definition of multiplication (9.6) that $f(x)g(x)$ cannot have degree greater than $n + m$. Moreover, since S is an integral domain and we know that $a_n \neq 0$ and $b_m \neq 0$, it follows that the coefficient a_nb_m of x^{n+m} is not zero, and 9.11 follows at once.

Lemma 9.10 assures us that if $f(x)$ and $g(x)$ are nonzero elements of $S[x]$, with S an integral domain, then the element $f(x)g(x)$ of $S[x]$ has a degree and therefore is not zero. Hence, $S[x]$ is also an integral domain. Since $S \subset S[x]$, it is trivial that if $S[x]$ is an integral domain, then S must be an integral domain. We have thus completed the proof of Theorem 9.7.

The familiar property 9.11 is not necessarily true if S is not an integral domain since, in the above proof, a_nb_m might be zero without either factor being zero. For example, let $S = \mathbf{Z}_6$. If $f(x) = [1] + [2]x$ and $g(x) = [2] + [4]x + [3]x^2$, then $\deg f(x) = 1$ and $\deg g(x) = 2$. However, $f(x)g(x) = [2] + [2]x + [5]x^2$ and $\deg f(x)g(x) = 2$. In this case,

$$\deg\,(f(x)g(x)) < \deg f(x) + \deg g(x)\,.$$

In this section we have introduced polynomials in *one* indeterminate x. However, this procedure can easily be generalized as follows. If S is a commutative ring with unity, then the polynomial ring $S[x]$ is a commutative ring with unity. If now y is another indeterminate, we may as above construct a ring $(S[x])[y]$ consisting of polynomials in y with coefficients in the ring $S[x]$. It is easy to verify that the elements of this new ring can also be expressed as polynomials in x with coefficients in the ring $S[y]$; in other words, that the rings $(S[x])[y]$ and $(S[y])[x]$

are identical. Accordingly, we may denote this ring by $S[x, y]$ and call its elements polynomials in the indeterminates x and y. A double application of Theorem 9.7(iii) then assures us that $S[x, y]$ is an integral domain if and only if S is an integral domain. These statements may be extended in an obvious way to polynomials in any finite number of indeterminates. However, for the most part we shall study polynomials in just one indeterminate.

9.2 THE SUBSTITUTION PROCESS

In defining the polynomial ring $S[x]$, where S is a commutative ring with unity, we have emphasized that x is not to be considered as an element of S. However, if $f(x) = a_0 + a_1x + \cdots + a_nx^n$ is an element of $S[x]$ and $s \in S$, let us define

9.12 $$f(s) = a_0 + a_1s + \cdots + a_ns^n.$$

It follows that $f(s)$ is a uniquely determined element of S associated with the polynomial $f(x)$ and the element s of S. Now the importance of this "substitution process" stems from the fact that our definitions of addition and multiplication in $S[x]$ have the same form as though x were an element of S. Let us state this fact more precisely in terms of the mapping $\theta: S[x] \to S$ defined by

9.13 $$f(x)\theta = f(s), \qquad f(x) \in S[x].$$

We may emphasize that in this mapping we are thinking of s as being a fixed element of S. Different elements s of S would, of course, lead to different mappings of $S[x]$ into S. When we said above that addition and multiplication of polynomials were defined "as though x were an element of S," what we really meant was that the operations of addition and multiplication are preserved under the mapping θ, that is, that θ is a homomorphism. Actually, the mapping θ is a homomorphism of $S[x]$ *onto* S (no matter what element s of S is used) since if $a_0 \in S$ and $f(x) = a_0$, clearly $f(x)\theta = a_0$.

We are frequently interested in considering elements r of S such that $f(r) = 0$, and so we make the following definition.

9.14 Definition. If $f(x) \in S[x]$ and $r \in S$ such that $f(r) = 0$, we say that r is a *root* of the polynomial $f(x)$.*

* In elementary algebra, r is usually said to be a root of the *equation* $f(x) = 0$, in which case x is thought of as an unknown number. However, this is not consistent with the definitions of the preceding section, and we shall continue to write $f(x) = 0$ to mean that $f(x)$ is the zero polynomial.

In later sections we shall obtain various results about roots of polynomials. However, in order to obtain results of a familiar nature, we shall find it necessary to make some additional restrictions on the ring S. In particular, we shall frequently assume that S is a *field.* As an example to show what may happen if we do not restrict the ring of coefficients, let T be the ring of all subsets of a given set (Example 10 of Section 2.3), and $T[x]$ the ring of polynomials in the indeterminate x with coefficients in T. Since $a^2 = a$ for every element a of T, it is clear that the polynomial $x^2 - x$ of $T[x]$ has as a root *every* element of T. We thus have an example of a polynomial of degree 2 that has more than two roots (if the given set has more than one element). In the next section we shall see that this cannot happen in case the ring of coefficients is restricted to be a field.

EXERCISES

1. Prove the distributive laws in $S[x]$.

2. If S is a commutative ring with unity, verify that the set of all polynomials of $S[x]$ with zero constant terms is a subring of $S[x]$.

3. Verify that the set of all polynomials of $S[x]$ with the property that all odd powers of x have zero coefficients is a subring of $S[x]$. Is the same true if the word "odd" is replaced by the word "even"?

4. If $\mathbf{Z}$ is the ring of integers and x an indeterminate, let $(\mathbf{Z}[x])^+$ be the subset of $\mathbf{Z}[x]$ consisting of those nonzero polynomials which have as leading coefficient a *positive* integer. Show that the set $(\mathbf{Z}[x])^+$ has all the properties required in 3.4, and hence that $\mathbf{Z}[x]$ is an ordered integral domain.

5. Generalize the preceding exercise by showing that if D is an ordered integral domain, then the polynomial ring $D[x]$ is also an ordered integral domain.

6. Let $h(x)$ be the element $5x^2 - 3x + 4$ of $\mathbf{Z}_6[x]$. (Here we are writing 5, -3 and 4 in place of the more cumbersome $[5]$, $[-3]$, and $[4]$.) By simply trying all the elements of $\mathbf{Z}_6$, find all roots of $h(x)$ in $\mathbf{Z}_6$.

7. If $g(x)$ is the element $x^7 - x$ of $\mathbf{Z}_7[x]$, verify that all elements of $\mathbf{Z}_7$ are roots of $g(x)$.

8. If m is a positive integer, how many polynomials are there of degree m over the ring $\mathbf{Z}_n$ of integers modulo n?

9. Let S and T be commutative rings, each with a unity, and suppose that

$\theta: S \to T$ is a given homomorphism of S onto T. If a mapping $\phi: S[x] \to T[x]$ is defined by

$$(a_0 + a_1x + \cdots + a_nx^n)\phi = a_0\theta + (a_1\theta)x + \cdots + (a_n\theta)x^n,$$

prove that ϕ is a homomorphism of $S[x]$ onto $T[x]$.

10. If $f(x) = g(x)h(x)$, where these are elements of $\mathbf{Z}[x]$, and every coefficient of $f(x)$ is divisible by the prime p, prove that every coefficient of $g(x)$ is divisible by p or every coefficient of $h(x)$ is divisible by p. [Hint: Use the preceding exercise with $S = \mathbf{Z}$, $T = \mathbf{Z}_p$, and $\theta: \mathbf{Z} \to \mathbf{Z}_p$ as defined near the end of Section 4.7. Then consider what $[f(x)]\phi = 0$ implies about the polynomial $f(x)$.]

11. If

$$f(x) = \sum_{i=0} a_ix^i$$

is a polynomial over a commutative ring S, let us define the *derivative* $f'(x)$ of $f(x)$ as follows:

$$f'(x) = \sum_{i=1} ia_ix^{i-1}.$$

Prove that

$$[f(x) + g(x)]' = f'(x) + g'(x)$$

and that

$$[f(x)g(x)]' = f(x)g'(x) + f'(x)g(x).$$

9.3 DIVISORS AND THE DIVISION ALGORITHM

In this and the next two sections we shall study polynomials with coefficients in an arbitrary *field* F. We then know, by Theorem 9.7, that $F[x]$ is necessarily an integral domain. The following definition is essentially a restatement of Definition 4.1 as applied to the integral domain $F[x]$ instead of the integral domain $\mathbf{Z}$.

9.15 Definition. Let $F[x]$ be the ring of polynomials in the indeterminate x over an arbitrary field F. If $f(x)$, $g(x) \in F[x]$, $g(x)$ is said to be a *divisor* (or *factor*) of $f(x)$ if there exists $h(x) \in F[x]$ such that $f(x) = g(x)h(x)$. If $g(x)$ is a divisor of $f(x)$, we say also that $f(x)$ is *divisible* by $g(x)$ or that $f(x)$ is a *multiple* of $g(x)$.

It follows immediately from this definition that if c is a nonzero element of F (that is, a polynomial of $F[x]$ of degree zero), then c is a divisor of every element $f(x)$ of $F[x]$. For, since c has a multiplicative inverse c^{-1} in F, we can write $f(x) = c(c^{-1}f(x))$, and this shows that c is a divisor of $f(x)$.

As suggested by the notation used in the case of integers, we shall sometimes write $g(x)|f(x)$ to indicate that $g(x)$ divides (that is, is a divisor of) $f(x)$.

It is important to observe that if $f(x) = g(x)h(x)$, then also $f(x) = (cg(x))(c^{-1}h(x))$, where c is any nonzero element of F. That is, if $g(x)|f(x)$, then $(cg(x))|f(x)$ for every nonzero element c of F.

The following result plays just as important a role in the study of divisibility in $F(x)$ as the corresponding result (4.4) does in establishing properties of the integers.

9.16 Division Algorithm. *If $f(x), g(x) \in F[x]$ with $g(x) \neq 0$, there exist unique elements $q(x)$ and $r(x)$ of $F[x]$ such that*

9.17 $$f(x) = q(x)g(x) + r(x), \quad r(x) = 0 \text{ or } \deg r(x) < \deg g(x).$$

We may recall that the zero polynomial has no degree and this fact explains the form of the condition which $r(x)$ is required to satisfy.

PROOF. If $f(x)$ and $g(x)$ are given polynomials, the polynomials $q(x)$ and $r(x)$ can easily be computed by the usual process of long division. The existence of such polynomials therefore seems almost obvious. However, we shall give a detailed proof of their existence, and for the moment leave aside the question of their uniqueness. Let us first dispose of two easy cases as follows.

(A) If $f(x) = 0$ or $\deg f(x) < \deg g(x)$, then 9.17 is trivially satisfied with $q(x) = 0$ and $r(x) = f(x)$.

(B) If $\deg g(x) = 0$, so that $g(x) = c$ with c a nonzero element of F, then $f(x) = (c^{-1}f(x))c$ and 9.17 holds with $q(x) = c^{-1}f(x)$ and $r(x) = 0$.

We are now ready to complete the proof by induction on the degree of $f(x)$. In this case, we shall use the form of the Induction Principle given in Exercise 9 of Section 3.3. If n is a positive integer, let S_n be the statement, "For every polynomial $f(x)$ of degree n and every nonzero polynomial $g(x)$, there exist polynomials $q(x)$ and $r(x)$ satisfying Equation 9.17." Let us now consider the statement S_1. By (A) and (B), we need only consider the case in which $\deg g(x) = 1$; that is, $g(x) = cx + d$, $c \neq 0$. Since $f(x) = ax + b$, $a \neq 0$, we can easily see that

9.18 $$f(x) = ac^{-1}(cx + d) + b - ac^{-1}d,$$

and 9.17 is satisfied with $q(x) = ac^{-1}$ and $r(x) = b - ac^{-1}d$. Hence, S_1 is true. Now suppose that k is a positive integer with the property that S_i is true for every positive integer $i \leq k$, and let us prove that S_{k+1} is true. Let $f(x) = ax^{k+1} + \cdots$, where $a \neq 0$, be a polynomial of degree $k + 1$ and let $g(x)$ be an entirely arbitrary polynomial. Cases (A) and (B) show that we may assume that $0 < \deg g(x) \leq k + 1$ since otherwise the existence of $q(x)$ and $r(x)$ satisfying 9.17 follows immediately. Suppose that $\deg g(x) = m$, hence that $g(x) = bx^m + \cdots$, with $b \neq 0$ and $0 < m \leq k + 1$. Now it is easily verified that

9.19 $$f(x) = b^{-1}ax^{k+1-m}g(x) + [f(x) - b^{-1}ax^{k+1-m}g(x)].$$

Perhaps we should point out that this equation is merely the result of taking one step in the usual long-division process of dividing $f(x)$ by $g(x)$. If we set $t(x) = f(x) - b^{-1}ax^{k+1-m}g(x)$, it is easy to see that the coefficient of x^{k+1} in $t(x)$ is zero, hence that $t(x) = 0$ or $\deg t(x) < k + 1$. By (A), or by the assumption that S_i is true for every positive integer $i \leq k$, we know that there exist polynomials $s(x)$ and $r(x)$, with $r(x) = 0$ or $\deg r(x) < \deg g(x)$, such that $t(x) = s(x)g(x) + r(x)$. Substituting in 9.19, we see that

$$f(x) = [b^{-1}ax^{k+1-m} + s(x)]g(x) + r(x),$$

and 9.17 is satisfied. Hence S_{k+1} is true, and it follows that S_n is true for every positive integer n. This completes the proof of the *existence* part of the Division Algorithm. The proof of the fact that $q(x)$ and $r(x)$ are *unique* will be left as an exercise.

Is is customary to call $q(x)$ and $r(x)$ satisfying 9.17 the *quotient* and the *remainder*, respectively, in the division of $f(x)$ by $g(x)$. Clearly, $f(x)$ is divisible by $g(x)$ if and only if the remainder in the division of $f(x)$ by $g(x)$ is zero.

A special case of the Division Algorithm of importance is that in which the divisor $g(x)$ is of the special form $x - c$, $c \in F$. In this case the remainder must be zero or have degree zero, that is, it is an element of F. We can thus write

$$f(x) = q(x)(x - c) + r, \qquad r \in F.$$

From this equation it is clear that $f(c) = r$, and hence that

$$f(x) = q(x)(x - c) + f(c).$$

The next two theorems then follow immediately.

9.20 Remainder Theorem. *If $f(x) \in F[x]$ and $c \in F$, the remainder in the division of $f(x)$ by $x - c$ is $f(c)$.*

9.21 Factor Theorem. *If $f(x) \in F[x]$ and $c \in F, f(x)$ is divisible by $x - c$ if and only if $f(c) = 0$, that is, if and only if c is a root of the polynomial $f(x)$.*

We shall now make use of the Factor Theorem to prove the following result.

9.22 Theorem. *Let F be a field and $f(x)$ an element of $F[x]$ of positive degree n and with leading coefficient a. If $c_1, c_2, \cdots, c_n$ are distinct elements of F, all of which are roots of $f(x)$, then*

9.23 $$f(x) = a(x - c_1)(x - c_2) \cdots (x - c_n).$$

PROOF. The proof of this theorem is by induction on the degree n of $f(x)$. If n is a positive integer, let S_n be the statement, "The statement of the theorem is true for every polynomial of degree n." We then wish to prove that S_n is true for every positive integer n. The truth of the statement S_1 follows quite easily. If $f(x)$ is of degree 1 and has leading coefficient a, then $f(x) = ax + b$, $a \neq 0$. If c_1 is a root of $f(x)$, we have $f(c_1) = 0$ or $ac_1 + b = 0$. Then $b = -ac_1$ and hence $f(x) = a(x - c_1)$, which is the desired form 9.23 in case $n = 1$.

Now let k be a positive integer such that S_k is true, and consider S_{k+1}. Accordingly, we let $f(x)$ be a polynomial of degree $k + 1$ with leading coefficient a, and let $c_1, c_2, \cdots, c_{k+1}$ be distinct roots of $f(x)$. Since c_1 is a root of $f(x)$, we have $f(c_1) = 0$ and by the Factor Theorem it follows that

9.24 $$f(x) = q(x)(x - c_1).$$

Now it is clear that deg $q(x) = k$ and the leading coefficient of $q(x)$ is a since a is the coefficient of x^{k+1} in $f(x)$. If c_i $(i \neq 1)$ is any other of the given roots of $f(x)$, it follows, using 9.24 and the fact that $f(c_i) = 0$, that

$$q(c_i)(c_i - c_1) = 0.$$

Since the c's are distinct, $c_i - c_1 \neq 0$ and therefore $q(c_i) = 0$. We have therefore shown that the polynomial $q(x)$ in 9.24 is of degree k, has leading coefficient a, and has $c_2, c_3, \cdots, c_{k+1}$ as distinct roots. Since S_k is assumed to be true, it follows that

$$q(x) = a(x - c_2)(x - c_3) \cdots (x - c_{k+1}).$$

Substituting this expression for $q(x)$ in 9.24, we get

$$f(x) = a(x - c_1)(x - c_2) \cdots (x - c_{k+1}).$$

Hence, S_{k+1} is true, and the Induction Principle assures us that S_n is true for every positive integer n. This completes the proof of the theorem.

We next establish the following corollary.

9.25 Corollary. *A polynomial $f(x)$ of degree n over a field F cannot have more than n distinct roots in F.*

PROOF. Since polynomials of degree zero have no roots, in verifying this corollary we may assume that $n \geq 1$. If $c_1, c_2, \cdots, c_n$ are distinct roots of $f(x)$, then $f(x)$ can be written in the form 9.23. Now let c be an arbitrary root of $f(x)$. Since $f(c) = 0$, it follows at once from 9.23 that

$$a(c - c_1)(c - c_2) \cdots (c - c_n) = 0.$$

Since $a \neq 0$, some one of the other factors must be zero, that is, $c = c_i$ for some i. Hence $c_1, c_2, \cdots, c_n$ are the *only* roots of $f(x)$, and $f(x)$ cannot have more than n distinct roots.

The next corollary is now a simple consequence of this one.

9.26 Corollary. *Let $g(x)$ and $h(x)$ be polynomials over a field F with the property that $g(s) = h(s)$ for every element s of F. If the number of elements in F exceeds the degrees of both $g(x)$ and $h(x)$, then necessarily $g(x) = h(x)$.*

PROOF. Let us set $f(x) = g(x) - h(x)$, and we then have that $f(s) = 0$ for every element s of F. If $f(x) \neq 0$, its degree can certainly not exceed the degrees of both $g(x)$ and $h(x)$, and hence $f(x)$ would have more distinct roots than its degree. Since, by the preceding corollary, this is impossible, we must have $f(x) = 0$. Hence $g(x) = h(x)$, as required.

The following example shows that this last result is not true without the restriction on the number of elements of the field F. Let F be the field $\mathbf{Z}_3$, a field of three elements. If $g(x) = x^3$ and $h(x) = x$, it is easy to show by direct substitution that $g(s) = h(s)$ for every element s of F, but $g(x)$ and $h(x)$ are not equal elements of the polynomial ring $F[x]$. See also Exercise 7 of the preceding set.

9.4 GREATEST COMMON DIVISOR

It has been pointed out earlier that if c is a nonzero element of the field F, then the element $f(x)$ of $F[x]$ is divisible by the element $g(x)$ of $F[x]$ if and only if $f(x)$ is divisible by $cg(x)$. By choosing c as the multiplicative inverse of the leading coefficient of $g(x)$, the polynomial $cg(x)$ will have the unity 1 of F as its leading coefficient. Hence, we will know *all* the divisors of $f(x)$ when we have determined all those divisors that have 1 as leading coefficient. The following definition makes it easy to refer to such polynomials.

9.27 Definition. A nonzero element of $F[x]$ is said to be a *monic* polynomial if its leading coefficient is the unity 1 of F.

The greatest common divisor of two elements of $F[x]$ may now be defined as follows.

9.28 Definition. The monic polynomial $d(x)$ of $F[x]$ is said to be the *greatest common divisor* (g.c.d) of the nonzero polynomials $f(x)$ and $g(x)$ of $F[x]$ if the following conditions are satisfied:

(i) $d(x)|f(x)$ and $d(x)|g(x)$.
(ii) If $h(x) \in F[x]$ such that $h(x)|f(x)$ and $h(x)|g(x)$, then $h(x)|d(x)$.

As in the case of integers, it is quite easy to verify that two polynomials cannot have more than one g.c.d. We proceed to outline a proof of the existence of the g.c.d. and to develop a method for actually computing the g.c.d. of two given polynomials. Inasmuch as the procedure follows quite closely the material of Section 4.3, we shall omit most of the details.

First, we make the following definition.

9.29 Definition. If $f(x), g(x) \in F[x]$, we say that a polynomial of the form

$$f(x)s(x) + g(x)t(x), \qquad s(x), t(x) \in F[x],$$

is a *linear combination* of $f(x)$ and $g(x)$.

The following theorem can now be established by a simple modification of the proof of Theorem 4.12.

9.30 Theorem. *If $f(x)$ and $g(x)$ are nonzero elements of $F[x]$, the monic polynomial of least degree which is expressible as a linear combination of $f(x)$ and $g(x)$ is the g.c.d. of $f(x)$ and $g(x)$. Hence, if $d(x)$ is the g.c.d. of $f(x)$ and $g(x)$, there exist elements $s_1(x)$ and $t_1(x)$ of $F[x]$ such that*

$$d(x) = f(x)s_1(x) + g(x)t_1(x),$$

and $d(x)$ is the monic polynomial of least degree which is expressible in this form.

In order to *compute* the g.c.d. of two nonzero polynomials $f(x)$ and $g(x)$ of $F[x]$, we use the Euclidean Algorithm as in the case of integers. By repeated use of the Division Algorithm we obtain the following sequence of equations, it being understood that $r_k(x)$ is the last nonzero remainder (and $r_k(x) = g(x)$ if $r(x) = 0$):

9.31

$$\begin{aligned}
f(x) &= q(x)g(x) + r(x), & \deg r(x) &< \deg g(x),\\
g(x) &= q_1(x)r(x) + r_1(x), & \deg r_1(x) &< \deg r(x),\\
r(x) &= q_2(x)r_1(x) + r_2(x), & \deg r_2(x) &< \deg r_1(x),\\
&\cdots\cdots\cdots\cdots & &\cdots\cdots\cdots\\
r_{k-2}(x) &= q_k(x)r_{k-1}(x) + r_k(x), & \deg r_k(x) &< \deg r_{k-1}(x),\\
r_{k-1}(x) &= q_{k+1}(x)r_k(x).
\end{aligned}$$

Now from these equations it follows that $r_k(x)$ is a divisor of both $f(x)$ and $g(x)$; also that any divisor of both $f(x)$ and $g(x)$ is a divisor of $r_k(x)$. If c is the leading coefficient of $r_k(x)$, then $c^{-1}r_k(x)$ also has these same properties and, moreover, it is a *monic* polynomial. We have therefore outlined a proof of the following result.

9.32 Theorem. *Let $r_k(x)$ be the last nonzero remainder in the Euclidean Algorithm as applied to the nonzero polynomials $f(x)$ and $g(x)$ of $F[x]$. If c is the leading coefficient of $r_k(x)$, then $c^{-1}r_k(x)$ is the g.c.d. of $f(x)$ and $g(x)$.*

In a numerical case, the actual calculations may often be simplified by the following observation. If $d(x)$ is the g.c.d. of $f(x)$ and $g(x)$, then also $d(x)$ is the g.c.d. of $af(x)$ and $bg(x)$, where a and b are nonzero elements of F. Hence, instead of the first of Equations 9.31, we might use the similar equation obtained by dividing $af(x)$ by $bg(x)$. In like manner, instead of the second equation we might work with $dg(x)$ and $er(x)$, where d and e are nonzero elements of F; and so on for the other equations. This modification will not affect the validity of the arguments used to show that $c^{-1}r_k(x)$ is the g.c.d. of $f(x)$ and $g(x)$, and may greatly simplify the work involved. Let us give an illustration by finding the g.c.d. of the polynomials

$$f(x) = x^3 + \tfrac{1}{2}x^2 + \tfrac{1}{3}x + \tfrac{1}{6}$$

and

$$g(x) = x^2 - \tfrac{1}{2}x - \tfrac{1}{2}$$

over the field $\mathbf{Q}$ of rational numbers. In order to avoid fractions, we divide $6f(x)$ by $2g(x)$, obtaining

$$6f(x) = (3x + 3)[2g(x)] + 8x + 4,$$

so that $r(x) = 8x + 4$. If we now divide $2g(x)$ by $r(x)/4$, we see that

$$2g(x) = (x - 1)[r(x)/4].$$

Since $r_1(x) = 0$, the g.c.d. of $f(x)$ and $g(x)$ is obtained from the last nonzero remainder, namely $8x + 4$, by multiplying it by the multiplicative inverse of its leading coefficient. Hence the g.c.d. of $f(x)$ and $g(x)$ is $x + \frac{1}{2}$.

It is sometimes convenient to use the following terminology, which is suggested by the corresponding definition for the integers.

9.33 Definition. Two nonzero elements $f(x)$ and $g(x)$ of $F[x]$ are said to be *relatively prime* if and only if their g.c.d. is 1.

EXERCISES

1. Complete the proof of the Division Algorithm by showing that the quotient and the remainder are unique.
2. If $f(x) \in F[x]$, show that $f(x)$ has as a factor a polynomial of $F[x]$ of degree one if and only if $f(x)$ has a root in F.
3. If F is the field $\mathbf{Z}_7$, use the result of Exercise 7 of the preceding set to show, without calculation, that in $F[x]$ we have

$$x^7 - x = x(x - 1)(x - 2)(x - 3)(x - 4)(x - 5)(x - 6).$$

4. State and prove a corresponding result for the field $\mathbf{Z}_p$, where p is an arbitrary prime. [Hint: Consider the multiplicative group of $\mathbf{Z}_p$.]
5. Prove Theorem 9.30.

6. Find the g.c.d. of each of the following pairs of polynomials over the field **Q** of rational numbers, and express it as a linear combination of the two polynomials:

(i) $2x^3 - 4x^2 + x - 2$ and $x^3 - x^2 - x - 2$,
(ii) $x^4 + x^3 + x^2 + x + 1$ and $x^3 - 1$,
(iii) $x^5 + x^4 + 2x^3 - x^2 - x - 2$ and $x^4 + 2x^3 + 5x^2 + 4x + 4$,
(iv) $x^3 - 2x^2 + x + 4$ and $x^2 + x + 1$.

7. Find the g.c.d. of each of the following pairs of polynomials over the indicated field, and express it as a linear combination of the two polynomials:

(i) $x^3 + 2x^2 + 3x + 2$ and $x^2 + 4$; field $\mathbf{Z}_5$,
(ii) $x^3 + (2i + 1)x^2 + ix + i + 1$ and $x^2 + (i - 1)x - 2i - 2$; field **C** of complex numbers,
(iii) $x^2 + (1 - \sqrt{2})x - \sqrt{2}$ and $x^2 - 2$; field **R** of real numbers,
(iv) $x^4 + x + 1$ and $x^2 + x + 1$; field $\mathbf{Z}_2$.

8. Let $f(x)$ and $g(x)$ be nonzero elements of $F[x]$, where F is a field. If the field F' is an extension of the field F, then $F[x] \subseteq F'[x]$ and we may also consider $f(x)$ and $g(x)$ to be elements of $F'[x]$. Show that the quotient and the remainder in the division of $f(x)$ by $g(x)$ are the same whether these polynomials are considered as elements of $F[x]$ or of $F'[x]$. In particular, conclude that if there exists an element $h(x)$ of $F'[x]$ such that $f(x) = g(x)h(x)$, then $h(x) \in F[x]$.

9. Verify that the Division Algorithm (9.16) remains true if the field F is replaced by an arbitrary commutative ring S with unity, provided only that $g(x)$ is required to have as leading coefficient an element of S with a multiplicative inverse in S.

10. By using the result of the preceding exercise, verify that the Factor Theorem and the Remainder Theorem are true if the field F is replaced by a commutative ring S with unity.

11. Give an example to show that Theorem 9.22 is not necessarily true if the field F is replaced by an arbitrary commutative ring S with unity. Where does the proof break down? Verify that the proof of this theorem will remain valid if F is replaced by an integral domain.

9.5 UNIQUE FACTORIZATION IN $F[x]$

We begin this section with the following definition.

9.34 Definition. A polynomial $p(x)$ of positive degree over a field F is said to be a *prime* (or *irreducible*) polynomial over F if it cannot be expressed as the product of two polynomials of positive degree over F.

If c is a nonzero element of the field F and $f(x) \in F[x]$, then we always have $f(x) = c^{-1}(cf(x))$, so that every polynomial of the form $cf(x)$ is a divisor of $f(x)$. It is easy to verify that a polynomial $f(x)$ of positive degree over F is a prime polynomial over F if and only if the *only* elements of $F[x]$ of positive degree that are divisors of $f(x)$ are of the form $cf(x)$, $c \neq 0$.

Since the degree of the product of two polynomials over F is the sum of the degrees of the factors, it follows at once from Definition 9.34 that *every element of $F[x]$ of the first degree is necessarily prime over F.*

We may emphasize that the possible divisors of $p(x)$ that are being considered in Definition 9.34 are those which are elements of $F[x]$; that is, they must have coefficients in F. For example, consider the polynomial $x^2 - 2$ over the field $\mathbf{Q}$ of rational numbers. Now $x^2 - 2$ cannot be factored into the product of two polynomials of the first degree in $\mathbf{Q}[x]$, and hence $x^2 - 2$ is a prime polynomial over $\mathbf{Q}$. However, if we should consider the same polynomial as a polynomial over the field $\mathbf{R}$ of real numbers, we find that it is *not* prime over $\mathbf{R}$ since we have the factorization $x^2 - 2 = (x - \sqrt{2})(x + \sqrt{2})$ with these factors of the first degree having coefficients in $\mathbf{R}$. As this example shows, the concept of a polynomial being a prime polynomial is relative to a specified field which contains the coefficients of the given polynomial.

In later sections we shall discuss prime polynomials over each of the familiar fields of elementary algebra. As for the finite fields of the form $\mathbf{Z}_p$, where p is a prime integer, we may here state without proof the following fact. For each prime p and each positive integer n, there exists at least one polynomial of degree n over the field $\mathbf{Z}_p$ which is prime over $\mathbf{Z}_p$.

The prime polynomials play essentially the same role in the factorization of an element of $F[x]$ as do the prime integers in the factorization of an integer. We therefore state without proof the following lemma and theorem, which are analogous to 4.18 and 4.20, respectively.

9.35 Lemma. *If $f(x)$ and $g(x)$ are nonzero polynomials over the field F and $p(x)$ is a prime polynomial over F such that $p(x)|(f(x)g(x))$, then $p(x)|f(x)$ or $p(x)|g(x)$.*

9.36 Theorem. *If $f(x)$ is a polynomial of positive degree over the field F and a is its leading coefficient, then there exist distinct monic prime polynomials $p_1(x), \cdots, p_k(x)$ $(k \geq 1)$ over F such that*

9.37 $$f(x) = a[p_1(x)]^{n_1}[p_2(x)]^{n_2} \cdots [p_k(x)]^{n_k},$$

where the n's are positive integers. Moreover, such a factorization is unique except for the order of the factors.

We may emphasize that the prime polynomials in 9.37 are restricted to be monic polynomials, and hence the leading coefficient of the right side is just a, which is given as the leading coefficient of $f(x)$. A proof of the above theorem can be given by a simple modification of the proof of the Fundamental Theorem of Arithmetic. Although Theorem 9.36 certainly has some theoretical significance, it is not so very useful from a computational point of view. For example, from the factorizations of two polynomials in the form 9.37 it is easy to write down their g.c.d. just as in the case of two integers. However, it is often very difficult to *find* the prime factors of a given polynomial and hence to write it in the form 9.37. Accordingly, it will usually be very much easier to apply the method of Section 9.4 to find the g.c.d. of two polynomials than to make use of Theorem 9.36.

As an important special case, one or more of the monic prime polynomials occurring in a factorization 9.37 of $f(x)$ may be of the first degree. In particular, the Factor Theorem (9.21) assures us that $x - c$, $c \in F$, is a factor of $f(x)$ if and only if $f(c) = 0$, that is, if and only if c is a root of the polynomial $f(x)$. The following definition introduces a terminology which is sometimes convenient.

9.38 Definition. The element c of F is said to be a root of *multiplicity* $m \geq 1$ of the polynomial $f(x)$ over F if $f(x)$ is divisible by $(x - c)^m$ but not by $(x - c)^{m+1}$. A root of multiplicity two is called a *double root.*

It follows that c is a root of $f(x)$ of multiplicity m if and only if in the factorization 9.37 of $f(x)$ one of the prime factors occurring is $x - c$ and, furthermore, it occurs with the exponent m.

EXERCISES

1. (a) Prove that a polynomial $f(x)$ of degree 2 or 3 over a field F is a prime polynomial over F if and only if the polynomial $f(x)$ has no root in F.

(b) Show, by means of an example, that a corresponding statement does not hold for polynomials of degree 4.

2. Determine whether or not each of the following polynomials is prime over each of the given fields. If it is not prime, factor it into a product of prime factors over each given field. As usual, **Q** is the field of rational numbers, **R** the field of real numbers, and **C** the field of complex numbers.

(a) $x^2 + x + 1$ over **Q**, **R**, and **C**;
(b) $x^2 + 2x - 1$ over **Q**, **R**, and **C**;
(c) $x^2 + 3x - 4$ over **Q**, **R**, and **C**;
(d) $x^3 + 2$ over **Q**, **R**, and **C**;
(e) $x^2 + x + 1$ over $\mathbf{Z}_2$, $\mathbf{Z}_3$, and $\mathbf{Z}_5$;
(f) $x^3 + x + 1$ over $\mathbf{Z}_2$, $\mathbf{Z}_5$, and $\mathbf{Z}_{11}$;
(g) $x^4 - 1$ over $\mathbf{Z}_{17}$;
(h) $x^3 + x^2 + 1$ over $\mathbf{Z}_{11}$;
(i) $x^2 + 15$ over **R** and **C**.

3. Find all prime polynomials of degree not more than 5 over the field $\mathbf{Z}_2$.

4. In each case the polynomial over the given field has as a root the specified element of the field. Find the multiplicity of this root and complete the factorization of the polynomial into prime factors over the given field.

(a) $x^4 + x^3 - 3x^2 - 5x - 2$ over **Q**, root -1;
(b) $x^4 + 1$ over $\mathbf{Z}_2$, root 1;
(c) $x^4 + 2x^2 + 1$ over **C**, root i;
(d) $x^4 + 6x^3 + 3x^2 + 6x + 2$ over $\mathbf{Z}_7$, root 4.

5. Prove that there are $(p^2 - p)/2$ monic quadratic polynomials which are prime over the field $\mathbf{Z}_p$. [Hint: First determine the number that are *not* prime.]

6. Prove that there are $(p^3 - p)/3$ monic cubic polynomials which are prime over the field $\mathbf{Z}_p$.

7. Suppose that $f(x)$ is a polynomial over the field F and let $f'(x)$ be the derivative of $f(x)$ as defined in Exercise 11 of Section 9.2. Prove each of the following:

(i) If an element c of F is a root of $f(x)$ of multiplicity greater than one, then c is also a root of the polynomial $f'(x)$.
(ii) If an element c of F is a root of $f(x)$ of multiplicity one, then c is not a root of $f'(x)$.
(iii) If $f(x)$ can be expressed as a product of elements of $F[x]$ of the first degree, then $f(x)$ and $f'(x)$ are relatively prime if and only if $f(x)$ has no root of multiplicity greater than one.

9.6 RATIONAL ROOTS OF A POLYNOMIAL OVER THE RATIONAL FIELD

If $f(x)$ is a polynomial of degree $n > 0$ over a field F, then clearly $f(x)$ and $cf(x)$ have the same roots for any nonzero element c of F. If, in particular, $f(x)$ has coefficients in the field **Q** of rational numbers and we choose c as the l.c.m. of the denominators of the coefficients of $f(x)$, then $cf(x)$ will have coefficients that are integers. In studying the roots of a polynomial with rational coefficients there is therefore no loss of generality in restricting attention to polynomials that have integral coefficients. We shall now prove the following theorem.

9.39 Theorem. *Let*

$$f(x) = a_nx^n + a_{n-1}x^{n-1} + \cdots + a_0, \qquad a_n \neq 0,$$

be a polynomial of positive degree n with coefficients that are integers. If r/s is a rational number, in lowest terms, which is a root of the polynomial $f(x)$, then r is a divisor of a_0 and s is a divisor of a_n.

We may recall that by saying that r/s is in lowest terms we mean that r and s are relatively prime integers and $s > 0$. However, the requirement that s be positive plays no role in the proof of this theorem.

PROOF. Since r/s is assumed to be a root of $f(x)$, we have that

$$a_n\left(\frac{r}{s}\right)^n + a_{n-1}\left(\frac{r}{s}\right)^{n-1} + \cdots + a_0 = 0.$$

If we multiply throughout by the nonzero integer s_n, we obtain

9.40 $$a_nr^n + a_{n-1}r^{n-1}s + \cdots + a_1rs^{n-1} + a_0s^n = 0.$$

By transposing the last term to the right side, this equation can be written in the form

$$(a_nr^{n-1} + a_{n-1}r^{n-2}s + \cdots + a_1s^{n-1})r = -a_0s^n.$$

Since all letters here represent integers, we see that the integer a_0s^n is divisible by the integer r. But we are given that r and s are relatively prime, and it therefore follows that a_0 is divisible by r.

By a similar argument, if in 9.40 we transpose a_nr^n to the other side, we can see that a_n is divisible by s.

As an example of the use of this theorem, let us find all rational roots of the polynomial

$$g(x) = 4x^5 + x^3 + x^2 - 3x + 1\,.$$

If r/s is a rational number, in lowest terms, which is a root of this polynomial, then r must be a divisor of 1 and s a positive divisor of 4. It follows that $r = \pm 1$, $s = 1, 2$, or 4; and we see that the only possible rational roots are the following: $1, \frac{1}{2}, \frac{1}{4}, -1, -\frac{1}{2}, -\frac{1}{4}$. It is easy to verify by direct calculation that $g(1) \neq 0, g(\frac{1}{2}) = 0, g(\frac{1}{4}) \neq 0, g(-1) = 0, g(-\frac{1}{2}) \neq 0$, and $g(-\frac{1}{4}) \neq 0$. Hence, $\frac{1}{2}$ and -1 are the *only* rational roots. If we divide $g(x)$ by $x - \frac{1}{2}$ and then divide the quotient by $x + 1$, we find that

$$g(x) = (x - \tfrac{1}{2})(x + 1)(4x^3 - 2x^2 + 4x - 2)\,.$$

Any root of this third degree factor is naturally a root of $g(x)$, so its only possible rational roots are therefore $\frac{1}{2}$ and -1. It is easy to verify that $\frac{1}{2}$ is a root and if we again divide by $x - \frac{1}{2}$, we can express $g(x)$ in the form

$$g(x) = (x - \tfrac{1}{2})^2(x + 1)(4x^2 + 4)$$

or

9.41 $$g(x) = 4(x - \tfrac{1}{2})^2(x + 1)(x^2 + 1)\,.$$

We see therefore that $\frac{1}{2}$ is a double root of $g(x)$. Since the quadratic polynomial $x^2 + 1$ has no rational root, it is a prime polynomial over **Q** and hence in 9.41 we have $g(x)$ expressed as a product of prime polynomials over **Q**. For that matter, the polynomial $x^2 + 1$ is prime over the field **R** of real numbers and so 9.41 also gives the factorization of $g(x)$ into prime polynomials over **R**.

EXERCISES

1. Complete the proof of Theorem 9.39 by showing that s is a divisor of a_n.
2. Prove the following corollary of Theorem 9.39. A rational root of a *monic* polynomial with coefficients that are integers is necessarily an integer which is a divisor of the constant term of the polynomial.
3. Find the factorization of the polynomial $g(x)$ of the example given above into prime factors over the field **C** of complex numbers.
4. Find all rational roots of each of the following polynomials over the rational field **Q**:

   (a) $3x^3 + 5x^2 + 5x + 2$,
   (b) $2x^4 - 11x^3 + 17x^2 - 11x + 15$.

(c) $x^5 - x^4 - x^3 - x^2 - x - 2$,
(d) $x^3 + x^2 - 2x - 3$,
(e) $6x^3 - 7x^2 - 35x + 6$,
(f) $x^5 + 5x^4 + 13x^3 + 19x^2 + 18x + 8$,
(g) $x^3 - (\frac{1}{5})x^2 - 4x + \frac{4}{5}$,
(h) $x^7 + x^6 + x^5 + x^4 + x^3 + x^2 + x + 1$.

5. Find all rational roots of each of the following polynomials over the rational field $\mathbf{Q}$, and factor each polynomial into a product of prime polynomials over $\mathbf{Q}$:

(a) $9x^4 + 6x^3 + 19x^2 + 12x + 2$,
(b) $x^5 - x^4 - 3x^3 + 6x^2 - 4x + 1$,
(c) $4x^4 + 20x^3 + 33x^2 + 20x + 4$,
(d) $2x^4 + 3x^3 + 4x + 6$.

6. Show that each of the following polynomials over $\mathbf{Q}$ has no rational root:

(a) $x^{1000} - x^{500} + x^{100} + x + 1$,
(b) $x^{12} - x^9 + x^6 - x^3 + 1$,
(c) $x^m + 2x^{m-1} - 2$, m a positive integer ≥ 2.

9.7 PRIME POLYNOMIALS OVER THE RATIONAL FIELD

It was pointed out in Section 9.5 that every polynomial of the first degree over a field F is necessarily a prime polynomial over F. Also, the first exercise at the end of that section asserts that a polynomial of degree 2 or 3 over F is a prime polynomial over F if and only if it has no root in F. If the field F is now taken to be the field $\mathbf{Q}$ of rational numbers, it is easy to apply Theorem 9.39 to find whether or not a polynomial of degree at most 3 is prime over $\mathbf{Q}$. For a polynomial of higher degree it may be exceedingly difficult to determine whether or not it is a prime. For example, a polynomial of degree 4 over $\mathbf{Q}$ may not have a rational root, and therefore may have no factor of the first degree over $\mathbf{Q}$, but may be a product of two prime polynomials of degree 2. In this section we shall give some rather special results, which will enable us to show that for *every* positive integer n, there exist polynomials of degree n that are prime over $\mathbf{Q}$.

We shall first prove two lemmas, the first of which is the following. As usual, $\mathbf{Z}[x]$ is the ring of polynomials in the indeterminate x with coefficients in the ring $\mathbf{Z}$ of integers.

9.42 Lemma. *Let $f(x)$, $g(x)$, and $h(x)$ be elements of the ring $\mathbf{Z}[x]$ such that $f(x) = g(x)h(x)$. If p is a prime integer which is a divisor of every coefficient of $f(x)$, then p is a divisor of every coefficient of $g(x)$ or a divisor of every coefficient of $h(x)$.*

PROOF. The proof of this lemma was stated as Exercise 10 at the end of Section 9.2, and one method of proof was suggested in a hint given there. We here indicate a more elementary proof which, however, does involve a little more calculation. Let us set

$$
\begin{aligned}
f(x) &= a_0 + a_1x + \cdots + a_nx^n, \\
g(x) &= b_0 + b_1x + \cdots + b_mx^m, \\
h(x) &= c_0 + c_1x + \cdots + c_kx^k.
\end{aligned}
$$

We are given that each coefficient a_i $(i = 0, 1, \cdots, n)$ is divisible by the prime p. Suppose now that $g(x)$ has at least one coefficient which is not divisible by p, and also that $h(x)$ has at least one coefficient which is not divisible by p, and let us seek a contradiction. To be more precise, let b_s be the *first* coefficient of $g(x)$, when $g(x)$ is written in increasing powers of x, that is not divisible by p; and let c_t be the *first* coefficient of $h(x)$ that is not divisible by p. Since $f(x) = g(x)h(x)$, by considering the coefficients of x^{s+t} on both sides of this equation, we find that

$$a_{s+t} = \cdots + b_{s-1}c_{t+1} + b_sc_t + b_{s+1}c_{t-1} + \cdots.$$

Now, by our choice of s and t, p is seen to be a divisor of every term on the right except the term b_sc_t. Since also p is a divisor of a_{s+t}, it follows that p is a divisor of b_sc_t. In view of the fact that p is a prime, this implies that p must be a divisor of b_s or a divisor of c_t. We have therefore obtained the desired contradiction. It follows that either $g(x)$ or $h(x)$ must have all coefficients divisible by p, and the proof is complete.

The following lemma, whose proof will be based on the preceding lemma, shows that a polynomial with *integral* coefficients is prime over the field $\mathbf{Q}$ if and only if it cannot be factored into a product of two polynomials of positive degree with *integral* coefficients. It will then be possible to prove that certain polynomials are prime over $\mathbf{Q}$ by making use of special properties of the integers.

9.43 Lemma. *Let $f(x)$ be an element of $\mathbf{Z}[x]$ such that $f(x) = g(x)h(x)$, where $g(x)$, $h(x) \in \mathbf{Q}[x]$. Then there exist polynomials $g'(x)$,*

$h'(x)$ of $\mathbf{Z}[x]$ having the same degrees as $g(x)$ and $h(x)$, respectively, such that $f(x) = g'(x)h'(x)$.

PROOF. Let k be the l.c.m. of the denominators of the coefficients of $g(x)$, so that $kg(x)$ has integral coefficients. Similarly, let l be an integer such that $lh(x)$ has integral coefficients. Since $f(x) = g(x)h(x)$, it follows that

9.44
$$klf(x) = g_1(x)h_1(x),$$

where $g_1(x)$ and $h_1(x)$ have integral coefficients. We may then apply the preceding lemma as follows. If p is a prime divisor of kl, it must be a divisor of all coefficients of $g_1(x)$ or of $h_1(x)$; hence p can be divided from both sides of Equation 9.44, and we still have polynomials with integral coefficients. By a repetition of this process, we can divide out every prime factor of kl and finally get $f(x) = g'(x)h'(x)$, where $g'(x)$ and $h'(x)$ have integral coefficients. It is almost trivial that $g'(x)$ has the same degree as $g(x)$, and also that $h'(x)$ has the same degree as $h(x)$. The proof is therefore complete.

We are now ready to prove the following theorem of Eisenstein.

9.45 Theorem. *Let $f(x) = a_0 + a_1x + \cdots + a_nx^n$ be a polynomial of positive degree n over the ring $\mathbf{Z}$ of integers, and p a prime integer such that $a_i \equiv 0 \pmod{p}$ for $i = 0, 1, \cdots, n - 1$; $a_n \not\equiv 0 \pmod{p}$ and $a_0 \not\equiv 0 \pmod{p^2}$. Then $f(x)$ is a prime polynomial over $\mathbf{Q}$.*

PROOF. The preceding lemma shows that we need only prove that $f(x)$ cannot be factored into a product of two factors of positive degree over $\mathbf{Z}$. Let us assume that

9.46
$$a_0 + a_1x + \cdots + a_nx^n = (b_0 + b_1x + \cdots + b_mx^m)(c_0 + c_1x + \cdots + c_kx^k),$$

where all these coefficients are integers, and clearly $m + k = n$. Since $a_0 = b_0c_0$, the fact that $a_0 \equiv 0 \pmod{p}$ but $a_0 \not\equiv 0 \pmod{p^2}$ shows that exactly one of the integers b_0 and c_0 is divisible by p. Suppose, for convenience of notation, that $c_0 \equiv 0 \pmod{p}$ and that $b_0 \not\equiv 0 \pmod{p}$. Now $a_n = b_mc_k$ and $a_n \not\equiv 0 \pmod{p}$; so $c_k \not\equiv 0 \pmod{p}$. Let s be chosen as the smallest positive integer such that $c_s \not\equiv 0 \pmod{p}$. From what we have just shown we know that there exists such an integer s and that $0 < s \leq k$. Now by a consideration of the coefficients of x^s on both sides of 9.46, we see that

$$a_s = b_0c_s + b_1c_{s-1} + \cdots,$$

and, in view of our choice of s, every term on the right with the single exception of b_0c_s is divisible by p. Moreover, $b_0 \not\equiv 0 \pmod{p}$ and $c_s \not\equiv 0 \pmod{p}$, so $a_s \not\equiv 0 \pmod{p}$. However, by our assumptions, the only coefficient of $f(x)$ that is not divisible by p is the leading coefficient a_n. Hence $s = n$, and therefore we must have $k = n$. This shows that in any factorization of $f(x)$ into a product of polynomials with integral coefficients, one of the factors must have degree n. It follows that $f(x)$ is necessarily a prime polynomial over $\mathbf{Q}$.

9.47 Corollary. *If n is an arbitrary positive integer, there exist polynomials of degree n over $\mathbf{Q}$ that are prime over $\mathbf{Q}$.*

PROOF. This result is easily established by examples. As an illustration, the polynomial $x^n - 2$ over $\mathbf{Q}$ satisfies all the conditions of the preceding theorem with $p = 2$. Hence $x^n - 2$ is a prime polynomial over $\mathbf{Q}$ for each positive integer n. In like manner, each of the following polynomials of degree n over $\mathbf{Q}$ is prime over $\mathbf{Q}$: $x^n + 2$, $x^n + 3$, $3x^n + 2x^{n-1} + 2x^{n-2} + \cdots + 2x + 2$, $x^n + 9x + 3$ $(n > 1)$. The reader will have no difficulty in constructing other examples.

Perhaps we should emphasize that we have not presented a general method for determining whether or not a given polynomial over $\mathbf{Q}$ is prime over $\mathbf{Q}$. This is a difficult problem, and we shall not discuss it further in this book.

9.8 POLYNOMIALS OVER THE REAL OR COMPLEX NUMBERS

In this section we shall discuss some properties of polynomials over the field $\mathbf{R}$ of real numbers or the field $\mathbf{C}$ of complex numbers. We begin with a few remarks, essentially established in elementary algebra, about quadratic polynomials; that is, polynomials of degree 2.

Let

$$g(x) = ax^2 + bx + c, \qquad a \neq 0,$$

be a quadratic polynomial with coefficients in the field $\mathbf{C}$. Then it is well known that the polynomial $g(x)$ has roots r_1 and r_2, where

9.48 $$r_1 = \frac{-b + \sqrt{b^2 - 4ac}}{2a}, \quad r_2 = \frac{-b - \sqrt{b^2 - 4ac}}{2a}.$$

We may point out that, by a special case of Theorem 6.20, every nonzero complex number has two square roots. Hence r_1 and r_2, given by 9.48, are complex numbers and it is easy to verify by direct calculation that

9.49 $$g(x) = a(x - r_1)(x - r_2).$$

Since these first-degree factors have coefficients in **C**, it is apparent that no quadratic polynomial over **C** is a prime polynomial over **C**.

It is customary to call $b^2 - 4ac$ the *discriminant* of the quadratic polynomial $ax^2 + bx + c$. For convenience, let us designate this discriminant by D.

From 9.48 it follows that $r_1 = r_2$ if and only if $D = 0$. However, the factorization 9.49 holds in any case, so $D = 0$ is a necessary and sufficient condition that the polynomial $g(x)$ have a double root.

Now let us assume that the quadratic polynomial $g(x)$ has *real* coefficients. Then the roots r_1 and r_2 will also be real if and only if $D \geq 0$, for only in this case will D have real square roots. The factorization 9.49 of $g(x)$ into factors of the first degree is therefore a factorization over **R** if and only if $D \geq 0$. If $D < 0$, $g(x)$ has no real root and $g(x)$ is therefore prime over **R**.

Let us summarize some of these observations in the following theorem.

9.50 Theorem. *No quadratic polynomial over the field* **C** *of complex numbers is prime over* **C**. *A quadratic polynomial over the field* **R** *of real numbers is prime over* **R** *if and only if its discriminant is negative.*

We have referred above to Theorem 6.20, where it was proved by use of the trigonometric form of a complex number that every nonzero complex number has n nth roots. It may be worth pointing out that the *square* roots of a complex number may also be computed by an algebraic process. As an illustration, let us seek the roots of the polynomial $x^2 + x - (1 + 3i)$ over **C**. By 9.48, these roots can immediately be written down in the form

9.51 $$\frac{-1 \pm \sqrt{5 + 12i}}{2}.$$

Now in order to express these roots in the usual form of complex numbers, we need to compute the square roots of $5 + 12i$. To do so, suppose that s and t are unknown real numbers such that $s + ti$ is a square root of $5 + 12i$. Thus we have

$$(s + ti)^2 = 5 + 12i,$$

or

$$s^2 - t^2 + 2sti = 5 + 12i.$$

In turn, this implies both of the following equations involving the real numbers s and t:

$$s^2 - t^2 = 5, \quad 2st = 12.$$

If we solve these two simultaneous equations by elementary methods and remember that s and t are real (so that $s^2 \geq 0$ and $t^2 \geq 0$), we find the solutions to be $s = 3$, $t = 2$ and $s = -3$, $t = -2$. Hence, the square roots of $5 + 12i$ are $\pm(3 + 2i)$. Substituting in 9.51, we find that the roots of the polynomial $x^2 + x - (1 + 3i)$ are $1 + i$ and $-(2 + i)$.

We have shown above that no quadratic polynomial is prime over **C**. Another special case of some interest is the following. Let us consider a polynomial of the form $ax^n + b$, where a and b are nonzero complex numbers and n is an arbitrary positive integer greater than 1. Since, by Theorem 6.20, the complex number $-b/a$ has n distinct nth roots and these are obviously roots of the polynomial $ax^n + b$, Theorem 9.22 asserts that this polynomial can be factored over **C** into a product of factors of the first degree. In particular, such a polynomial can never be prime over **C**.

The general theorem which we shall next state is partially suggested by the special cases already discussed. The theorem was first proved by the famous German mathematician Carl Friedrich Gauss (1777–1855), and is of such importance that it has often been called "The Fundamental Theorem of Algebra." Unfortunately, there is no really elementary proof of this theorem and we shall therefore have to omit the proof.

9.52 Theorem. *If $f(x)$ is an element of* **C**$[x]$ *of positive degree, there exists an element of* **C** *which is a root of the polynomial $f(x)$.*

If r is a complex number which is a root of the polynomial $f(x)$ of degree n over **C**, then in **C**$[x]$ we can use the Factor Theorem and write

$$f(x) = (x - r)f_1(x),$$

where $f_1(x)$ is of degree $n - 1$. It is then apparent from this observation and Theorem 9.36 that the preceding theorem can be expressed in either of the following alternate forms.

9.53 Theorem. *The* only *prime polynomials of* $\mathbf{C}[x]$ *are the polynomials of the first degree.*

9.54 Theorem. *If* $f(x)$ *is an element of* $\mathbf{C}[x]$ *of positive degree, then* $f(x)$ *is itself of the first degree or it can be factored in* $\mathbf{C}[x]$ *into a product of polynomials of the first degree.*

We next consider the question of which polynomials over the real field **R** are prime over **R**. Of course, the polynomials of the first degree are always prime, and we have shown in Theorem 9.50 that the quadratic polynomials over **R** that are prime over **R** are those with negative discriminant. A little later we shall prove that these are the only prime polynomials over **R**. First, however, we need a preliminary result, which is of some interest in itself.

Let

$$f(x) = a_n x^n + \cdots + a_1 x + a_0$$

be a polynomial of positive degree with real coefficients. Since $\mathbf{R} \subset \mathbf{C}$, then also $f(x) \in \mathbf{C}[x]$ and Theorem 9.52 states that there exists an element r of $\mathbf{C}$ such that $f(r) = 0$. We now want to make use of the concept of the conjugate of a complex number, introduced in Section 6.4. We recall that if $u = a + bi$ is a complex number, then the conjugate u^* of u is defined by: $u^* = a - bi$. It was shown that the mapping $u \rightarrow u^*$ is a one-one mapping of $\mathbf{C}$ onto $\mathbf{C}$, which preserves the operations of addition and multiplication. Now since $f(x)$ is assumed to have real coefficients and a real number is equal to its conjugate, it is not difficult to verify that

$$[f(r)]^* = a_n(r^*)^n + \cdots + a_1 r^* + a_0 = f(r^*).$$

But, since $f(r) = 0$, it follows that $[f(r)]^* = 0$ and therefore $f(r^*) = 0$. That is, r^* is also a root of the polynomial $f(x)$. This result we state as the following theorem.

9.55 Theorem. *If* r *is a complex number which is a root of the polynomial* $f(x)$ *with real coefficients, then the conjugate* r^* *of* r *is also a root of* $f(x)$.

If it happens that r is a real number, then $r^* = r$, and this theorem has no content. However, if r is not real, then r^* and r are distinct roots of $f(x)$. It follows that in $\mathbf{C}[x]$ we have

$$f(x) = (x - r)(x - r^*)f_1(x),$$

with the degree of $f_1(x)$ two less than the degree of $f(x)$. If $r = a + bi$, then $r^* = a - bi$, and a simple calculation shows that

$$(x - r)(x - r^*) = x^2 - 2ax + a^2 + b^2.$$

We can therefore write

9.56 $$f(x) = (x^2 - 2ax + a^2 + b^2)f_1(x),$$

and the quadratic factor on the right clearly has *real* coefficients. Since also $f(x)$ has real coefficients, it is easy to verify that $f_1(x)$ must have real coefficients (cf. Exercise 8, Section 9.4). It follows that 9.56 gives a factorization of $f(x)$ in $\mathbf{R}[x]$. If $\deg f(x) > 2$, $f(x)$ can therefore not be a prime polynomial over $\mathbf{R}$. This result, combined with Theorem 9.50, completes the proof of the following theorem.

9.57 Theorem. *The only polynomials of* $\mathbf{R}[x]$ *that are prime over* $\mathbf{R}$ *are the polynomials of the first degree and the quadratic polynomials with negative discriminant.*

Now a polynomial of *odd* degree clearly cannot be factored into a product of quadratic polynomials. Therefore, if a polynomial $f(x)$ of $\mathbf{R}[x]$ of odd degree is expressed as a product of prime polynomials over $\mathbf{R}$, at least one of these prime polynomials (in fact, an odd number of them) must be of the first degree. This implies that $f(x)$ has at least one real root, and the following is therefore an almost immediate consequence of the preceding theorem.

9.58 Corollary. *A polynomial with real coefficients and of odd degree necessarily has a real root.*

Except for quadratic polynomials, and polynomials of the special form $ax^n + b$, we have not given any indication as to how one might actually *find* the real or complex roots of a given polynomial. This is a difficult problem but some information can be found in texts on the "theory of equations." In particular, there do exist algebraic formulas for the roots of polynomials of degrees 3 or 4 with real or complex coefficients. Although these formulas are of great theoretical interest, they are not convenient to use in a numerical case. It is, however, not too difficult to develop methods of approximating the roots to any desired accuracy, and this is what is usually done in practical applications.

EXERCISES

1. Find the roots of each of the following polynomials and express each root in the standard form $a + bi$ of a complex number:

(a) $x^2 - (3i - 2)x - 5 - i$, (b) $x^2 + ix + 1$,
(c) $x^2 - (2 + i)x - 1 + 7i$, (d) $x^2 + x + 4$,
(e) $x^2 - x + 2 + \sqrt{2}i$, (f) $x^2 + 2x + i$.

2. Factor each of the following polynomials of $\mathbf{R}[x]$ into a product of prime polynomials over $\mathbf{R}$:

(a) $x^3 - 2x - 4$, (b) $x^3 - x^2 - 3x + 6$,
(c) $x^4 + 1$, (d) $x^4 + 2x^2 - 8$,
(e) $x^4 + x^3 + 2x^2 + x + 1$, (f) $x^5 + 1$.

Chapter 10

Ideals and Quotient Rings

In an early chapter we defined homomorphisms of rings and established a few simple properties of homomorphisms. Later on we introduced homomorphisms of groups and carried the theory far enough to show the importance of normal subgroups and quotient groups. We now return to the study of rings and bring that theory up to approximately the same level as was reached with groups. In particular, we shall introduce the concept of ideal in a ring as an analogue of normal subgroup of a group, and then define quotient rings and indicate their fundamental connection with homomorphisms. It will be seen that the ring $\mathbf{Z}_n$ of integers modulo n is an important special case of a quotient ring. We shall also study in some detail another important special case involving a polynomial ring $F[x]$ over a field. This will give us a method of constructing many new rings and fields in addition to those previously mentioned.

10.1 IDEALS

Let us begin with the following definition.

10.1 Definition. Let A be a subring of the ring R. Then

(i) A is said to be a *right ideal* in R if A is closed with respect to multiplication on the right by elements of R (if $a \in A$ and $r \in R$, then $ar \in A$).

(ii) A is said to be a *left ideal* in R if A is closed with respect to multiplication on the left by elements of R (if $a \in A$ and $r \in R$, then $ra \in A$).

(iii) A is said to be an *ideal* in R if it is both a right ideal in R and a left ideal in R (that is, it is closed with respect to multiplication on either side by elements of R).

We shall primarily be interested in ideals in a ring although the study of right (or left) ideals plays an important role in more advanced ring theory.

In any ring R, the subring consisting only of the zero element is clearly an ideal, and the entire ring R is also an ideal. These two ideals are often called trivial ideals. Another simple observation is that if R has a unity e and a right ideal A in R contains an element a with a multiplicative inverse, then $A = R$. For if $a \in A$ and $aa^{-1} = e$, then $e \in A$ and $ex = x \in A$ for every x in R. Clearly, a similar result holds for left ideals and ideals.

As a first example of an ideal (other than the trivial ones mentioned above), we may observe that the subring E of even integers in the ring $\mathbf{Z}$ is an ideal in $\mathbf{Z}$. This is true since the product of an even integer by an arbitrary integer is an even integer.

The subring $\mathbf{Z}$ of integers in the ring $\mathbf{Q}$ of rational numbers is not an ideal in $\mathbf{Q}$ since, for example, $3 \in \mathbf{Z}$ and $\frac{1}{2} \in \mathbf{Q}$, but $3 \cdot \frac{1}{2} \notin \mathbf{Z}$.

In order to give an example of a right ideal (or left ideal) which is not an ideal, we must clearly have a noncommutative ring. It is easy to verify that the subring $\{a, b\}$ of the ring K of Example 6 of Section 2.3 is a left ideal, but not a right ideal, in K. A more significant example is the following. Let W be the ring of all matrices of order two over the integers (Example 9 of Section 2.3). Then the set of all elements of W of the form

$$\begin{bmatrix} x & y \\ 0 & 0 \end{bmatrix}, \qquad x, y \in \mathbf{Z},$$

is a right ideal but not a left ideal; the set of all elements of W of the form

$$\begin{bmatrix} x & 0 \\ y & 0 \end{bmatrix}, \qquad x, y \in \mathbf{Z},$$

is a left ideal but not a right ideal; and the set of all elements of W of the form

$$\begin{bmatrix} x & y \\ z & t \end{bmatrix},$$

where x, y, z, and t are *even* integers, is an ideal in W. We leave to the reader the verification of these statements.

Throughout the rest of this section we shall consider *commutative* rings only, and hence there will be no distinction between ideals and right (or left) ideals. Moreover, for our purposes the most important case is that in which the ring has a unity. Accordingly, let S be a commutative ring with unity e. If $a \in S$, let

$$A = \{as \mid s \in S\},$$

and let us verify that A is an ideal in S. If s, $t \in S$, then $as + at = a(s + t)$ and $(as)t = a(st)$, so that A is clearly closed with respect to addition and with respect to multiplication by arbitrary elements of S. Moreover, $-(as) = a(-s) \in A$, and hence additive inverses of elements of A are elements of A. Thus, A is indeed an ideal in S. This conclusion holds whether or not S has a unity, but the presence of a unity now assures us that $a \in A$ since $a = ae$. It is customary to denote this ideal A by (a). As an example of this notation, the ideal of all even integers in the ring $\mathbf{Z}$ would be denoted by (2). Similarly, the ideal consisting of all multiples of 3 would be denoted by (3), and so on.

Some convenient terminology is introduced in the following definition.

10.2 Definition. Let S be a commutative ring with unity. If $a \in S$, an ideal of the form

$$(a) = \{as \mid s \in S\}$$

is called a *principal ideal*. It is also called the *principal ideal generated by a*.

The next theorem shows that in certain important rings there are no ideals except principal ideals.

10.3 Theorem. *Every ideal is a principal ideal*

(*i*) *in the ring* $\mathbf{Z}$ *of integers,*
(*ii*) *in a polynomial ring* $F[x]$ *over a field.*

PROOF. We shall prove the first case and outline the proof for the other case. Suppose that K is an ideal in the ring $\mathbf{Z}$. If K is the zero ideal, then K is the principal ideal (0) generated by the zero of $\mathbf{Z}$. If $K \neq (0)$, then K must contain positive integers. (Why?) Suppose that m is the least positive integer in K, and let k denote an arbitrary element of K. By the Division Algorithm, we may write $k = qm + r$, where $0 \leq r < m$. Now $qm \in K$ since $m \in K$, and $r = k - qm$ is an element of K. Since $r < m$ and m is the least positive integer in K, we conclude that $r = 0$ and $k = qm$. It follows that $K = (m)$, completing this part of the proof.*

Suppose now that K is an ideal in the polynomial ring $F[x]$, where F is a field with unity 1, and again we may assume that $K \neq (0)$. If K contains a nonzero element r of F, then $1 = rr^{-1} \in K$ and $K = F[x] = (r)$. If $K \neq (0)$ and $K \neq F[x]$, let $f(x)$ be a polynomial of least degree in the ideal K. Then $f(x)$ has positive degree and the method of proof of the previous case will show that $K = (f(x))$. We leave the details as an exercise.

Let us observe that the rings $\mathbf{Z}$ and $F[x]$ are quite special rings, and the result just proved for these rings does not hold in general. We shall now emphasize this fact by giving an example of an ideal which is not a principal ideal. For our ring we take $\mathbf{Z}[x]$, the ring of polynomials over $\mathbf{Z}$. It may be verified that the set T of all elements of $\mathbf{Z}[x]$ with *even* constant terms is an ideal in $\mathbf{Z}[x]$. However, we shall show that T is not a principal ideal. Suppose, on the contrary, that $T = (f(x))$. Since $2 \in T$, we must have $2 = f(x)g(x)$ for some element $g(x)$ of $\mathbf{Z}[x]$. It follows that both $f(x)$ and $g(x)$ are of degree zero, that is, they are integers. Now $f(x) \neq \pm 1$ since this would imply that $(f(x)) = \mathbf{Z}[x]$, which is false. Accordingly, we must have $f(x) = \pm 2$, $g(x) = \pm 1$. However, there are many elements of T (for example, $x + 2$) which are not multiples of ± 2. We have obtained a contradiction, and T is therefore not a principal ideal in $\mathbf{Z}[x]$.

EXERCISES

1. If A and B are ideals (right ideals, left ideals) in a ring R, prove that $A \cap B$ is an ideal (right ideal, left ideal) in R. Generalize to any number of ideals.

2. If A and B are ideals (right ideals, left ideals) in a ring R, let us define

$$A + B = \{a + b \mid a \in A, b \in B\}.$$

Prove that $A + B$ is an ideal (right ideal, left ideal) in R and that $A \subseteq A + B$ and $B \subseteq A + B$.

* This argument should seem familiar since it is essentially the proof (Theorem 7.30) that every subgroup of a cyclic group is cyclic. In fact, we have really proved that the ideals in $\mathbf{Z}$ coincide with the cyclic subgroups of the additive group of $\mathbf{Z}$.

3. Let s and t be nonzero integers with d as their g.c.d. and m as their l.c.m. Prove that in the ring $\mathbf{Z}$, $(s) \cap (t) = (m)$ and $(s) + (t) = (d)$.
4. Find integers s and t such that $(s) \cup (t)$ is not an ideal in $\mathbf{Z}$.
5. Complete the proof of Theorem 10.3 for the ring $F[x]$.
6. Verify that a *field* F has only the two trivial ideals (0) and F.
7. Verify that the set of all matrices of the form

$$\begin{bmatrix} x - 2y & y \\ 2x - 4y & 2y \end{bmatrix}, \qquad x, y \in \mathbf{Z},$$

is a right ideal in the ring of all matrices of order two over $\mathbf{Z}$.
8. An element a of a ring R is said to be *nilpotent* if $a^n = 0$ for some positive integer n (which may depend on a). Prove that the set of all nilpotent elements in a commutative ring R is an ideal in R.
9. Let a and b be elements of a commutative ring with unity. If a has a multiplicative inverse and b is nilpotent, prove that $a + b$ has a multiplicative inverse. [Hint: If $b^2 = 0$, then $(a + b)^{-1} = a^{-1} - a^{-2}b$. What if $b^3 = 0$? In general, what if $b^n = 0$?]
10. Let U be an infinite set (for example, the set $\mathbf{Z}^+$), and let R be the ring of all subsets of U. Prove that the set S of all elements of R consisting of *finite* subsets of U is an ideal in R. Prove that S is not a principal ideal in R.

10.2 QUOTIENT RINGS

Let us first make a few additional observations about homomorphisms of rings. Since a ring is an abelian group with respect to the operation of addition, and zero is the identity of this group, the following definition is at least partly suggested by the corresponding definition for groups.

10.4 Definition. If $\theta: R \to S$ is a homomorphism of the ring R into the ring S, the set of all elements a of R such that $a\theta = 0$ is called the *kernel* of the homomorphism θ, and denoted by $\ker \theta$.

Corresponding to Theorem 7.21 for groups, we have the following theorem for rings.

10.5 Theorem. *If $\theta: R \to S$ is a homomorphism of the ring R into the ring S, then $\ker \theta$ is an ideal in R. Moreover, θ is an isomorphism if and only if $\ker \theta = \{0\}$.*

PROOF. If $a \in \ker\theta$ and $r \in R$, then

$$(ar)\theta = (a\theta)(r\theta) = 0(r\theta) = 0,$$

and hence $ar \in \ker\theta$. Similarly, $ra \in \ker\theta$. Moreover, if $a, b \in \ker\theta$, then

$$(a + b)\theta = a\theta + b\theta = 0 + 0 = 0,$$

and $a + b \in \ker\theta$. Finally, if $a \in \ker\theta$, Theorem 2.33(ii) assures us that $(-a)\theta = -(a\theta) = 0$, and $-a \in \ker\theta$. We have thus shown that $\ker\theta$ is an ideal in R.

If θ is a one-one mapping and $a \in \ker\theta$, then $a\theta = 0\theta$ and we must have $a = 0$; hence $\ker\theta = \{0\}$. To go the other way, suppose that $\ker\theta = \{0\}$ and that $a, b \in R$ such that $a\theta = b\theta$. Then $(a - b)\theta = 0$ and this implies that $a - b = 0$. It follows that the mapping is a one-one mapping and therefore an isomorphism. This completes the proof of the theorem.

Now let K be an ideal in an arbitrary ring R. The additive group of K is clearly a subgroup of the additive group of R. We wish to consider cosets with respect to this subgroup. A typical such coset is of the form

$$a + K = \{a + k \mid k \in K\},$$

where a is an element of R. From our previous study of cosets, we know that $c + K = d + K$ if and only if $c \in d + K$, that is, if and only if $c - d \in K$. Moreover, we also know that addition of cosets is well-defined as follows:

10.6 $$(a + K) + (b + K) = (a + b) + K, \qquad a, b \in R.$$

So far we have only used the fact that the additive group of K is a subgroup of the additive group of R. Now, however, since K is an ideal we can show that multiplication of cosets is well-defined by

10.7 $$(a + K)(b + K) = ab + K, \qquad a, b \in R.$$

Suppose that $a + K = a_1 + K$ and that $b + K = b_1 + K$. Thus there exist elements $k, k' \in K$ such that $a = a_1 + k$ and $b = b_1 + k'$. It follows that

$$\begin{aligned} ab &= (a_1 + k)(b_1 + k') \\ &= a_1b_1 + a_1k' + kb_1 + kk' \\ &= a_1b_1 + k'', \end{aligned}$$

where $k'' = a_1k' + kb_1 + kk'$ is an element of K. Hence $ab + K = a_1b_1 + K$, and multiplication of cosets is well-defined by 10.7.

We now have the following result (cf. Theorem 7.53 for groups).

10.8 Theorem. *Let K be an ideal in the ring R. With respect to addition and multiplication of cosets defined by 10.6 and 10.7, the set of all cosets of K in R is a ring, usually called the* quotient ring *of R by K, and denoted by R/K. Moreover, the mapping $\theta: R \to R/K$ defined by $a\theta = a + K$, $a \in R$, is a homomorphism of R onto the ring R/K, with kernel K.*

The proof that we get a ring is straightforward and we leave it as an exercise. However, it may be well to observe here that the zero of the ring R/K is the coset $0 + K = K$. Moreover, if R has a unity e, then R/K has the unity $e + K$.

The last statement of the theorem is an almost immediate consequence of the definitions 10.6 and 10.7 of addition and multiplication in the ring R/K.

We now give an important example of a quotient ring with which we are already familiar. Let n be a positive integer and let us consider the principal ideal (n) in the ring $\mathbf{Z}$. A coset $a + (n)$ of the ideal (n) in $\mathbf{Z}$ consists of all integers x expressible in the form $a + nt$ for some $t \in \mathbf{Z}$. Thus, $x \in a + (n)$ if and only if $x \equiv a \pmod{n}$. In the notation of equivalence sets modulo n, introduced in Section 4.7, we see therefore that $a + (n) = [a]$. Moreover, our definitions of addition and multiplication of cosets coincide with the previous definitions of addition and multiplication of equivalence sets modulo n. Thus the ring $\mathbf{Z}_n$ of integers modulo n is, in the language and notation of the present section, precisely the quotient ring $\mathbf{Z}/(n)$. Some additional examples of quotient rings will be presented in the following section.

10.3 QUOTIENT RINGS $F[x]/(s(x))$

We now consider the ring $F[x]$ of all polynomials in an indeterminate x over a *field* F, and let $s(x)$ be a fixed element of $F[x]$ of *positive* degree. Moreover, for convenience in writing, we shall sometimes let $S = (s(x))$; that is, S is the ideal in $F[x]$ consisting of all polynomials of the form $s(x)g(x)$, $g(x) \in F[x]$. Our purpose is to study the quotient ring $F[x]/(s(x))$. An element of this ring is a coset of the form

$$f(x) + S, \qquad f(x) \in F[x],$$

where $S = (s(x))$. We shall first prove the following theorem.

10.9 Theorem. *The ring $F[x]/(s(x))$ is a field if and only if $s(x)$ is a prime polynomial over F. In any case, the ring $F[x]/(s(x))$ contains a subring isomorphic to the field F.*

PROOF. Suppose, first, that $s(x)$ is not prime over F. Then there exist elements $s_1(x)$ and $s_2(x)$ of $F[x]$ of positive degrees such that $s(x) = s_1(x)s_2(x)$. Since $\deg s_1(x) < \deg s(x)$, $s_1(x)$ cannot be divisible by $s(x)$ and therefore $s_1(x) \notin S$. Similarly, $s_2(x) \notin S$. It follows that neither $s_1(x) + S$ nor $s_2(x) + S$ is the zero S of $F[x]/S$. However,

$$(s_1(x) + S)(s_2(x) + S) = s_1(x)s_2(x) + S = s(x) + S = S,$$

since $s(x) \in S$. Hence a product of two nonzero elements is the zero element of $F[x]/S$. This shows that $F[x]/S$ is not even an integral domain, and therefore certainly not a field.

Next, suppose that $s(x)$ is prime over F and let $f(x) + S$ be a nonzero element of $F[x]/S$, that is, $f(x) \notin S$. Thus $f(x)$ is not divisible by $s(x)$ and since $s(x)$ is prime over $F[x]$, this implies that $f(x)$ and $s(x)$ are relatively prime. By Theorem 9.30, there must then exist elements $h(x)$ and $k(x)$ of $F[x]$ such that

$$f(x)h(x) + s(x)k(x) = 1,$$

where 1 is the unity of F. Since $s(x)k(x) \in S$, this implies that

$$(f(x) + S)(h(x) + S) = 1 + S.$$

However, $1 + S$ is the unity of $F[x]/S$, and this shows that the arbitrary nonzero element $f(x) + S$ of $F[x]/S$ has a multiplicative inverse $h(x) + S$. Hence $F[x]/S$ is indeed a field. The part of the theorem proved so far closely parallels corresponding results for the ring of integers modulo n.

We now proceed to the proof of the last sentence of the theorem. Let F' be the set of all elements of $F[x]/S$ of the form $a + S$, where $a \in F$. Since $S = (s(x))$ and $s(x)$ has positive degree, S contains no nonzero element of F and we see that

$$a + S = b + S, \qquad a, b \in F$$

if and only if $a - b \in S$, that is, if and only if $a = b$. This shows that the mapping $a \rightarrow a + S$ of F into F' is one-one, and it is clearly an onto mapping. The proof of the theorem is completed by verifying that F' is a field and that this mapping is an isomorphism of F onto F'. We leave the details as an exercise.

In the future we shall often find it convenient to identify F' with F, that is, we merely change the notation by writing a in place of $a + S$, where $a \in F$, whenever it is clear from the context that we are working in the ring $F[x]/S$.

We proceed to point out a convenient, and somewhat more explicit, way to specify the elements of the ring $F[x]/S$. As soon as we have done this, we shall give a number of examples that may help to clarify the material of this and of the preceding section.

We have assumed that the fixed polynomial $s(x)$ of $F[x]$ has positive degree, say k. If c is a nonzero element of F, then a polynomial is a multiple of $s(x)$ if and only if it is a multiple of $cs(x)$, that is, $(s(x)) = (cs(x))$. Accordingly, there is no loss of generality in assuming that $s(x)$ is a monic polynomial since this would only involve choosing c to be the multiplicative inverse of the leading coefficient of $s(x)$. We shall henceforth assume that $s(x)$ is monic and, for later convenience, we choose the notation

10.10 $$s(x) = x^k - s_{k-1}x^{k-1} - \cdots - s_1x - s_0,$$

the coefficients being elements of F. Now let $f(x) + S$ be an element of $F[x]/S$. By the Division Algorithm, we have $f(x) = q(x)s(x) + r(x)$, where $r(x) = 0$ or $\deg r(x) < k$. Since $q(x)s(x) \in S$, it follows that

$$f(x) + S = r(x) + S.$$

Thus every element of $F[x]/S$ can be expressed in the form

10.11 $$a_0 + a_1x + \cdots + a_{k-1}x^{k-1} + S, \qquad \text{each } a_i \in F.$$

We shall next show that every element is *uniquely* expressible in this form. Let $g(x) = a_0 + a_1x + \cdots + a_{k-1}x^{k-1}$ and $h(x) = b_0 + b_1x + \cdots + b_{k-1}x_k{}^{-1}$, and suppose that $g(x) + S = h(x) + S$. Then $g(x) - h(x) \in S$ and since S contains no polynomial of degree less than k, we conclude that $g(x) - h(x) = 0$, or $g(x) = h(x)$. Thus each element of $F[x]/S$ can be expressed in exactly one way in the form 10.11.

We proceed to simplify our notation by introducing a simple symbol for the particular element $x + S$ of $F[x]/S$. Let us agree to set

$$j = x + S.$$

Then $j^2 = (x + S)(x + S) = x^2 + S$ and, more generally, for each positive integer m we have

$$j^m = x^m + S.$$

Now, for example, let us consider an element of the form $(a + bx + cx^2) + S$, where $a, b, c \in F$. Clearly,

$$\begin{aligned}(a + bx + cx^2) + S &= (a + S) + (b + S)(x + S) + (c + S)(x^2 + S)\\ &= (a + S) + (b + S)j + (c + S)j^2.\end{aligned}$$

We agreed above to write a for $a + S$, $a \in F$, so using this notation we have

$$(a + bx + cx^2) + S = a + bj + cj^2.$$

By generalizing this argument, we see that an element $f(x) + S$ of $F[x]/S$ may be written in the simple form $f(j)$. In particular, from 10.11 it follows that the elements of $F[x]/S$ are uniquely expressible in the form

10.12 $$a_0 + a_1 j + \cdots + a_{k-1} j^{k-1}, \qquad \text{each } a_i \in F.$$

Since $s(x) \in S$, $s(x) + S$ is the zero of $F[x]/S$, and we have $s(j) = s(x) + S = 0$. Thus, in particular, *the element j is a root of the polynomial $s(x)$* (although j is in the ring $F[x]/S$, not in F). Accordingly, we can operate with the elements 10.12 of $F[x]/S$ by merely considering j to be a symbol such that $s(j) = 0$. The sum of two elements of the form 10.12 is immediately an element of the same form. The product of two elements can be expressed in the form 10.12 by multiplying out in the usual way and then replacing each power of j higher than the $(k - 1)$st by the element of the form 10.12 to which it is equal. Since $s(j) = 0$, we have from 10.10 that

10.13 $$j^k = s_0 + s_1 j + \cdots + s_{k-1} j^{k-1},$$

and the right side is of the form 10.12. To compute j^{k+1}, we multiply the preceding equation by j and obtain

$$j^{k+1} = s_0 j + s_1 j^2 + \cdots + s_{k-2} j^{k-1} + s_{k-1} j^k.$$

Now the right side is not of the form 10.12, but we can get it into this form by substituting for j^k from 10.13 and collecting coefficients of the different powers of j. In this way we obtain

10.14 $$j^{k+1} = s_{k-1}s_0 + (s_0 + s_{k-1}s_1)j + \cdots + (s_{k-2} + s_{k-1}^2)j^{k-1}.$$

We could proceed in this way to compute higher powers of j. However, the general formulas are not very useful since it is much easier to apply the method directly in any specific case. The following examples will illustrate how this is done.

Example 1: Let F be the field $\mathbf{R}$ of real numbers, and let $s(x)$ be the polynomial $x^2 + 1$. Since $x^2 + 1$ is prime over $\mathbf{R}$, we know by Theorem 10.9 that $\mathbf{R}[x]/(x^2 + 1)$ is a field. We proceed to describe this field in some detail.

As a special case of the general discussion above, we may observe that since $x^2 + 1$ is of the second degree, every element of the field $\mathbf{R}[x]/(x^2 + 1)$ is uniquely expressible in the form

10.15 $$a + bj, \qquad a, b \in \mathbf{R},$$

where $j^2 + 1 = 0$. Let us now characterize the field $\mathbf{R}[x]/(x^2 + 1)$ by specifying the operations of addition and multiplication of the elements 10.15. Addition is trivial since

10.16 $$(a + bj) + (c + dj) = (a + c) + (b + d)j,$$

which is immediately of the form 10.15. As for multiplication, we have

$$(a + bj)(c + dj) = ac + (ad + bc)j + bdj^2,$$

or, replacing j^2 by -1,

10.17 $$(a + bj)(c + dj) = (ac - bd) + (ad + bc)j.$$

We have thus shown how to express the product of two elements of the form 10.15 in the same form. The field $\mathbf{R}[x]/(x^2 + 1)$ can now be simply characterized as the field with elements 10.15 and with addition and multiplication given by 10.16 and 10.17, respectively.

Except for an almost trivial difference in notation, the field we have just constructed coincides with the field $\mathbf{C}$ of complex numbers as introduced in Chapter 6. More precisely, the field $\mathbf{R}[x]/(x^2 + 1)$ is isomorphic to the field $\mathbf{C}$ under the mapping

$$a + bj \rightarrow a + bi, \qquad a, b \in \mathbf{R}.$$

Hence we have now given another method of constructing the field of complex numbers from the field of real numbers.

Example 2: Let F be the field $\mathbf{Z}_2$ of integers modulo 2, whose two elements we shall now write as 0 and 1; and let $s(x) = x^2 + x + 1$. Since neither of the elements of $\mathbf{Z}_2$ is a root of this quadratic polynomial, it follows that this polynomial is prime over $\mathbf{Z}_2$. Hence $\mathbf{Z}_2[x]/(x^2 + x + 1)$ is a field which, for simplicity, we shall designate by F^*. Since also in this example $s(x)$ is a quadratic polynomial, it follows as in the preceding example that the elements of F^* are uniquely expressible in the form

10.18 $$a + bj, \qquad a, b \in \mathbf{Z}_2.$$

As usual, addition of two of these elements is carried out in an almost trivial way as follows:

10.19 $$(a + bj) + (c + dj) = (a + c) + (b + d)j.$$

As for multiplication, we have as in the previous example

$$(a + bj)(c + dj) = ac + (ad + bc)j + bdj^2.$$

However, since $s(j) = 0$, in this example we have $j^2 + j + 1 = 0$, or $j^2 = -j - 1$. Since our coefficients are from the field $\mathbf{Z}_2$, we can just as well write $j^2 = j + 1$. Replacing j^2 by $j + 1$ in the above expression for $(a + bj)(c + dj)$, we obtain

10.20 $$(a + bj)(c + dj) = (ac + bd) + (ad + bc + bd)j$$

as the general formula for the product of two elements of F^*. Since there are only two elements of $\mathbf{Z}_2$, and therefore only two choices for a and b in 10.18, we see that F^* has only the four elements $0, 1, j, 1 + j$. Using 10.19 and 10.20 or, better still, simply carrying out the calculations in each case, we can construct the following addition and multiplication tables for F^*.

$(+)$	0	1	j	$1+j$
0	0	1	j	$1+j$
1	1	0	$1+j$	j
j	j	$1+j$	0	1
$1+j$	$1+j$	j	1	0

$(\cdot)$	0	1	j	$1+j$
0	0	0	0	0
1	0	1	j	$1+j$
j	0	j	$1+j$	1
$1+j$	0	$1+j$	1	j

The reader may verify that this field F^* of four elements is isomorphic to the ring of Example 7 of Section 2.3.

Example 3: Let F be the field $\mathbf{Z}_2$, as in the preceding example, but let $s(x) = x^2$. In this case, $s(x)$ is certainly not prime over $\mathbf{Z}_2$. For convenience, let us denote the ring $\mathbf{Z}_2[x]/(x^2)$ by T. We know then, by Theorem 10.9, that T is not a field. The elements of T are again of the form

$$a + bj, \qquad a, b \in \mathbf{Z}_2,$$

only this time $j^2 = 0$. Using this fact, we can obtain the following addition and multiplication tables for T.

$(+)$	0	1	j	$1+j$
0	0	1	j	$1+j$
1	1	0	$1+j$	j
j	j	$1+j$	0	1
$1+j$	$1+j$	j	1	0

$(\cdot)$	0	1	j	$1+j$
0	0	0	0	0
1	0	1	j	$1+j$
j	0	j	0	j
$1+j$	0	$1+j$	j	1

The addition table coincides with the addition table of Example 2, but the multiplication table is different, as it would have to be since T is not a field.

Example 4: Let $s(x)$ be the polynomial $x^3 + x^2 + 1$ over the field $\mathbf{Q}$ of rational numbers, and let us next consider the ring $\mathbf{Q}[x]/(x^3 + x^2 + 1)$, which we shall denote by U. It is easy to verify that $x^3 + x^2 + 1$ has no rational root and, since it is of degree 3, it must therefore be prime over $\mathbf{Q}$. Hence, U is a field. Since the degree of $s(x)$ is 3 in this case, it follows, as a special case of 10.12, that the elements of U are uniquely expressible in the form

10.21 $$a + bj + cj^2, \qquad a, b, c \in \mathbf{Q}.$$

Let us consider the product of two of these elements as an illustration of the procedure by which we obtained 10.13 and 10.14. Since $s(j) = 0$ we have that $j^3 + j^2 + 1 = 0$, or

10.22 $$j^3 = -1 - j^2.$$

If we multiply this equation by j and substitute $-1 - j^2$ for j^3, we get

$$j^4 = -j - j^3 = -j - (-1 - j^2);$$

that is,

10.23 $$j^4 = 1 - j + j^2.$$

The product of two elements 10.21 can now be computed by using the distributive laws and then substituting for j^3 and j^4 from 10.22 and 10.23. If we do so, we finally obtain

10.24 $$\begin{aligned}(a + bj + cj^2)(d + ej + fj^2) &= ad - bf - ce + cf \\ &\quad + (ae + bd - cf)j + (af + be + cd - bf - ce + cf)j^2.\end{aligned}$$

The field U can therefore be characterized as the field with elements 10.21, with multiplication given by 10.24, and addition carried out in the obvious way. It should perhaps be remarked that the formula 10.24 is obviously too complicated to be of much practical use. For example, it would be extremely difficult to find the multiplicative inverse of a given element of U by using this formula for the product of two elements. Instead, one would use the method of proof of Theorem 10.9 in order to carry out such a calculation.

There is one further theorem which is implicit in what we have done, and which is of sufficient importance to warrant an explicit statement as follows.

10.25 Theorem. *If F is a field and $f(x)$ is an arbitrary element of $F[x]$ of positive degree, there exists an extension F' of the field F such that $f(x)$ has a root in F'.*

PROOF. If $f(x)$ has a root in F, then the result is trivial with $F' = F$. Otherwise, there exists a factor $s(x)$ of $f(x)$, which is of degree at least 2 and is prime over F. Then let us set $F' = F[x]/(s(x))$. We know that the field F' is an extension of the field F, and we may therefore also consider $f(x)$ to be a polynomial over F'. Moreover, using the notation in which the element $x + (s(x))$ of F' is denoted by j, we have that $s(j) = 0$. That is, the element j of F' is a root of the polynomial $s(x)$ and therefore also of $f(x)$. This completes the proof.

The following result is an easy consequence of the theorem just established.

10.26 Corollary. *If F is a field and $f(x)$ is an element of $F[x]$ of positive degree, there exists an extension F^* of the field F with the property that $f(x)$ factors in $F^*[x]$ into factors of the first degree.*

PROOF. Using the notation of the proof of the theorem, the Factor Theorem assures us that in $F'[x]$ the polynomial $f(x)$ has $x - j$ as a factor. If $f(x)$ does not factor entirely into factors of the first degree over F'; that is, if $f(x)$ contains a prime factor over F' which is of degree at least 2, the process can be repeated by constructing a field F'' which contains F' and in which $f(x)$ has another root. It is clear that by a continuation of this process, there exists a field F^* containing F such that $f(x)$ factors in $F^*[x]$ into factors of the first degree, and the corollary is proved.

EXERCISES

1. Find a homomorphism of the ring $\mathbf{Z}_{12}$ onto the ring $\mathbf{Z}_4$. What is the kernel of this homomorphism?
2. Let R be the ring of all matrices of the form

$$\begin{bmatrix} a & 0 \\ b & a \end{bmatrix}, \qquad a, b \in \mathbf{Z}.$$

Find a homomorphism of R onto $\mathbf{Z}$ and verify by direct calculation that the kernel K of this homomorphism is an ideal in R.

3. For the given field F and the given polynomial $s(x)$ over F, construct a multiplication table for the ring $F[x]/(s(x))$. Which of these rings are fields?

(a) $F = \mathbf{Z}_2$, $s(x) = x^2 + 1$;
(b) $F = \mathbf{Z}_3$, $s(x) = x^2 + 1$;
(c) $F = \mathbf{Z}_3$, $s(x) = x^2 + x + 1$;
(d) $F = \mathbf{Z}_2$, $s(x) = x^3 + x + 1$;
(e) $F = \mathbf{Z}_2$, $s(x) = x^3 + x^2 + 1$.

4. Determine all positive integers n with $1 < n \leq 7$ such that $\mathbf{Z}_n[x]/(x^2 + x + 1)$ is a field.

5. Discuss the field $\mathbf{Q}[x]/(x^2 - 2)$ and verify that it is isomorphic to the field of all real numbers of the form $a + b\sqrt{2}$, where $a, b \in \mathbf{Q}$.

6. Discuss the field $\mathbf{Q}[x]/(x^3 - 2)$, and describe a field of real numbers to which it is isomorphic.

7. It was stated earlier that for each positive integer n and each positive prime p, there exists a polynomial of degree n with coefficients in the field $\mathbf{Z}_p$ which is prime over this field. Use this fact to show that there exists a field with p^n elements. (See Exercise 16 below.)

8. In each case describe a field in which the given polynomial over the specified field has a root. In particular, give a general formula for the product of two elements of the field you describe.

(a) $x^3 + x + 1$ over $\mathbf{Q}$,
(b) $x^3 + x^2 + x + 2$ over $\mathbf{Q}$,
(c) $x^2 - x + 1$ over $\mathbf{R}$,
(d) $x^3 + x + 1$ over $\mathbf{Z}_2$.

9. (a) Compute the multiplicative inverse of $1 + j + j^2$ in the field U of Example 4 above, and check by use of Formula 10.24.
(b) Verify that in $U[x]$,

$$x^3 + x^2 + 1 = (x - j)(x^2 + (1 + j)x + j + j^2).$$

10. If R is a commutative ring and N is the ideal in R consisting of all nilpotent elements of R (Exercise 8 of preceding set), show that the quotient ring R/N contains no nilpotent element except the zero.

11. An ideal P in a commutative ring R is said to be a *prime ideal* if whenever $a, b \in R$ such that $ab \in P$, then $a \in P$ or $b \in P$. Determine all prime ideals in the ring $\mathbf{Z}$ and in the ring $F[x]$, where F is a field.

12. Prove: If R is a commutative ring with unity, an ideal $P \neq R$ in R is a prime ideal if and only if R/P is an integral domain.

13. In the notation of Exercise 2 above, show that the ring R/K is isomorphic to the ring $\mathbf{Z}$.

14. If A is a set of elements of a ring R and $\theta: R \to S$ is a homomorphism of R onto S, then $A\theta$ is the set of elements of S that are images of elements of A under the mapping θ so that, in particular, $R\theta = S$. Using the notation introduced in Exercise 14 at the end of Section 7.7, if U is a set of elements of S, we define $U\theta^{-1} = \{a \mid a \in R, a\theta \in U\}$. Prove each of the following:

(i) If A is a right ideal (left ideal, ideal) in the ring R, then $A\theta$ is a right ideal (left ideal, ideal) in the ring S.

(ii) If U is a right ideal (left ideal, ideal) in the ring S, then $U\theta^{-1}$ is a right ideal (left ideal, ideal) in the ring R and it contains ker θ.

(iii) The mapping $A \to A\theta$ is a one-one mapping of the set of all right ideals (left ideals, ideals) in R which contain ker θ onto the set of all right ideals (left ideals, ideals) in S.

(iv) If R is commutative, an ideal P in R which contains ker θ is a prime ideal in R if and only if $P\theta$ is a prime ideal in S.

15. Let p be a positive prime and n a positive integer, and consider the polynomial $f(x) = x^{p^n} - x$ as a polynomial over the field $\mathbf{Z}_p$. By Corollary 10.26, there exists an extension F of $\mathbf{Z}_p$ such that $f(x)$ factors in $F[x]$ into factors of the first degree. Prove that $f(x)$ has p^n distinct roots in F. [Hint: See Exercise 6 of Section 9.5.]

16. In the notation of the preceding exercise, prove that the p^n distinct roots of $f(x)$ in F are the elements of a subfield of F, and hence conclude that there exists a field with p^n elements. [Hint: See Exercise 8 of Section 5.3.]

10.4 THE FUNDAMENTAL THEOREM ON RING HOMOMORPHISMS

We conclude this chapter by presenting the following theorem which corresponds to Theorem 7.55 for groups.

10.27 Theorem. *Let $\phi: R \to S$ be a homomorphism of the ring R onto the ring S, with kernel K. Then K is an ideal in R, and S is isomorphic to the quotient ring R/K. More precisely, the mapping $\alpha: R/K \to S$ defined by*

10.28
$$(a + K)\alpha = a\phi, \qquad a \in R,$$

is an isomorphism of R/K onto S.

PROOF. We have already shown that K is an ideal in R. Moreover, by a suitable change in notation (addition replacing multiplication), the corresponding proof for groups will show that 10.28

actually defines a mapping of R/K onto S. We leave it as an exercise to show that α indeed is a homomorphism. If $a + K$ is in ker α, then $(a + K)\alpha = a\phi = 0$ and $a \in \ker \phi$. But $\ker \phi = K$, so $a \in K$ and $a + K = K$. Hence ker α consists only of the zero K of R/K and α is an isomorphism, as we wished to show.

This theorem may be interpreted as stating that if isomorphic rings are not considered as different, the *only* homomorphic images of a ring R are the quotient rings R/K, where K is an ideal in R.

EXERCISES

1. Prove in detail, giving reasons for each step, that 10.28 actually defines a mapping of R/K onto S, and that it is a homomorphism.
2. Show that for each positive integer n, there is exactly one ring (not counting isomorphic rings as different) with n elements which is a homomorphic image of the ring $\mathbf{Z}$ of integers.
3. Prove: If $\theta: F \to S$ is a homomorphism of a *field* F onto a ring S with more than one element, then θ is an isomorphism and S is a field.
4. Let A and B be ideals in a ring R and define $A + B$ as in Exercise 2 of Section 10.1. Verify that B is an ideal in the ring $A + B$ and that $A \cap B$ is an ideal in the ring A. Then prove that the quotient ring $A/(A \cap B)$ is isomorphic to the quotient ring $(A + B)/B$. [Hint: The mapping $a \to a + B$, $a \in A$ is a homomorphism of A onto $(A + B)/B$. What is its kernel?]

NOTES AND REFERENCES

For further topics in rings and ideals see the books [24] through [27] of the Bibliography and other references given in these books.

Chapter 11

Factorization in Integral Domains

In the integral domain **Z** and in each integral domain $F[x]$, where F is a field, we have found that we have in a certain sense unique factorization into prime factors. The purpose of this brief chapter is to introduce some preliminary notation and definitions and then to consider whether such a theorem holds in other integral domains. We shall give an example to show that there are integral domains in which one does *not* have unique factorization. The main part of the chapter will consist of sketching a proof that unique factorization does hold in an integral domain in which every ideal is principal—which we know to be true (by Theorem 10.3) for the integers and for the polynomials over a field.

11.1 INTEGRAL DOMAINS IN GENERAL

We shall here give some definitions and make some preliminary observations about divisibility in an arbitrary integral domain D. Thus D is a commutative ring with unity in which the cancellation law of multiplication holds for all nonzero elements.

Throughout this section D will be an integral domain with unity e.

11.1 Definition. An element of D which has a multiplicative inverse in D may be called a *unit* of D.

The word "unit" is not to be confused with the word "unity," although the unity of D is certainly one of its units.

We shall find it convenient to denote by U the set of all units of D. Since $e \in U$, $U \neq \varnothing$. If $a \in U$, then $a^{-1} \in U$ since a^{-1} has the multiplicative inverse a. It is also known that if c and d have multiplicative inverses c^{-1} and d^{-1}, respectively (in any ring with unity), then cd has the multiplicative inverse $d^{-1}c^{-1}$. Hence U is closed with respect to multiplication, and we conclude from Theorem 7.4 that U is a group with multiplication as the operation.

Just as in the case of integers or of polynomials, we make the following definition.

11.2 Definition. If $a, b \in D$ with $b \neq 0$, b is said to be a *divisor* (or *factor*) of a if there exists an element a_1 of D such that $a = a_1 b$. If b is a divisor of a, we say that b *divides* a or that a *is divisible by* b or that a *is a multiple of* b.

We shall often write $b|a$ to indicate that b divides a, and use $b \nmid a$ to indicate that b does not divide a.

We may observe that *the units are divisors of all elements of* D, since if $a \in D$ and $u \in U$, we have $a = auu^{-1}$.

Let us now prove the following simple, but important, result.

11.3 Lemma. *An element c of D is a unit if and only if $c|e$.*

PROOF. This follows by the following observations. First, if $c|e$, then $e = cx$ for some x in D and thus $c^{-1} = x$ and $c \in U$. Conversely, if $c \in U$, then $cc^{-1} = e$ and therefore $c|e$.

11.4 Definition. If $a = bu$, where $a, b \in D$ and $u \in U$, we say that a and b are *associates*, and each is said to be *an associate of* the other.

As suggested by the language just used, if a and b are associates, the relation between them is mutual. For if $a = bu$ with $u \in U$, then $b = au^{-1}$ and $u^{-1} \in U$.

We leave it to the reader to verify that "is an associate of" is an equivalence relation on the set D. Moreover, the associates of the unity e of D are precisely the units of D.

11.5 Lemma. *Nonzero elements a and b of an integral domain D are associates if and only if $a|b$ and $b|a$.*

PROOF. If $a|b$ and $b|a$, then there exist elements x and y of D such that $b = ax$ and $a = by$. Thus $a = axy$ and $e = xy$ since $a \neq 0$. It follows from Lemma 11.3 that x and y are units, and therefore a and b are associates. Conversely, if a and b are associates, $a = ub$ for some $u \in U$, and hence $b|a$. A similar argument shows that $a|b$, and the proof is complete.

Let us illustrate these various concepts by using the familiar integral domains **Z** and $F[x]$, where F is a field.

The units of **Z** are 1 and -1, so that $U = \{1, -1\}$ is a group of order 2. Each nonzero integer a has exactly two associates, namely a and $-a$.

In the polynomial domain $F[x]$, the units are the nonzero elements of F, that is, the polynomials of degree zero. In this case $U = \{c \mid c \in F, c \neq 0\}$, and this group will have infinite order unless F is a finite field. In any case, U is the multiplicative group of the field F. The associates of an element $f(x)$ of $F[x]$ are the elements $cf(x)$, where c is a nonzero element of F.

We next make the following definition.

11.6 Definition. An element a of an integral domain D is said to be *irreducible* if it is not a unit and its only divisors are units and associates of a.

In **Z** and in $F[x]$, where F is a field, we have previously used the word "prime" to describe an element with the property stated in Definition 11.6. However, the terminology introduced here is quite often used when studying an arbitrary integral domain. Then a prime element is defined in the following way.

11.7 Definition. An element p of an integral domain D is said to be *prime* if it has the property that if a and b are elements of D such that $p|(ab)$, then $p|a$ or $p|b$.

As to the relationship between these two concepts, we first establish the following result.

11.8 Lemma. *A prime of an integral domain is necessarily irreducible.*

PROOF. To prove this, let p be a prime of the integral domain D and suppose that $x|p$. Thus, there exists $y \in D$ such that $p = xy$.

Since $p|(xy)$ and p is assumed to be a prime, we have that $p|x$ or $p|y$. If $p|x$, Lemma 11.5 shows that x is an associate of p. If $p|y$, then $y = y_1p$ for some $y_1 \in D$. It follows that $p = xy_1p$ and $xy_1 = e$, thus showing that x is a unit. We have shown that a divisor of p is an associate of p or is a unit, and hence p is irreducible and the proof is complete.

This result, together with Lemma 4.18* and Lemma 9.35, shows that in the integral domain **Z** and in a polynomial ring over a field, the property of being irreducible (Definition 11.6) is equivalent to the property of being prime (Definition 11.7). We shall presently give an example of an integral domain in which there are some elements which are irreducible but are not prime, so the two concepts are not equivalent in all integral domains.

It will be convenient to make another definition as follows.

11.9 Definition. An integral domain D is said to be a *unique factorization domain* (U.F.D.) if it has both of the following properties:

(i) Every nonzero element a of D which is not a unit is an irreducible or can be expressed as a product of a finite number of irreducibles.

(ii) If $a = p_1p_2 \cdots p_r$ and $a = q_1q_2 \cdots q_s$, where all the p's and the q's are irreducibles, then $r = s$ and by a proper choice of notation, q_i and p_i are associates for each i.

Theorem 4.20 shows that **Z** is a U.F.D., and Theorem 9.36 shows that $F[x]$ also is a U.F.D. In the next section we shall see that each of these theorems may be considered to be a special case of a more general result.

Before proceeding, let us introduce and discuss in some detail an illuminating example of an integral domain which is quite different from these previously mentioned. We begin by defining a set $\mathbf{Z}[\sqrt{-5}]$ of complex numbers as follows:

$$\mathbf{Z}[\sqrt{-5}] = \{a + b\sqrt{-5} \mid a, b \in \mathbf{Z}\}.$$

It is easily verified that this set is a subring of the field of complex numbers, and it contains the unity since $1 = 1 + 0\sqrt{-5}$. This is

* We may emphasize that in these lemmas, the word "prime" is used for what is now being called "irreducible."

sufficient to show that $\mathbf{Z}[\sqrt{-5}]$ is actually an integral domain, and we proceed to discuss this integral domain.

Many of the properties of the integral domain $\mathbf{Z}[\sqrt{-5}]$ can be obtained by associating with each of its elements an element of $\mathbf{Z}$ in the following way. If $x = a + b\sqrt{-5}$ is an element of $\mathbf{Z}[\sqrt{-5}]$, we define the *norm* of x, which we indicate by $N(x)$, by setting $N(x) = a^2 + 5b^2$. Since a and b are integers, $N(x)$ is an integer, necessarily nonnegative. The most important property of the norm is the following, whose verification we leave to the reader to establish by direct calculation: If $x, y \in \mathbf{Z}[\sqrt{-5}]$, then

11.10 $$N(xy) = N(x) \cdot N(y).$$

Let us now determine the units of $\mathbf{Z}[\sqrt{-5}]$. If $a + b\sqrt{-5}$ is a unit, there exists an element $c + d\sqrt{-5}$ such that

$$(a + b\sqrt{-5})(c + d\sqrt{-5}) = 1.$$

Now taking norms of both sides of this equation and using 11.10, we see that $(a^2 + 5b^2)(c^2 + 5d^2) = N(1) = 1$. Since $a^2 + 5b^2$ and $c^2 + 5d^2$ are nonnegative integers, we must have $a^2 + 5b^2 = 1$ and $c^2 + 5d^2 = 1$. It follows that $b = 0$ and $a = \pm 1$, so that $a + b\sqrt{-5} = \pm 1$. Thus, the only units of $\mathbf{Z}[\sqrt{-5}]$ are ± 1. Incidentally, we have shown that the units are precisely those elements having norm 1.

As an illustration of how one can show the irreducibility of certain elements of $\mathbf{Z}[\sqrt{-5}]$, let us verify that 2 is irreducible. Suppose, on the contrary, that $2 = (a + b\sqrt{-5})(c + d\sqrt{-5})$. Then taking norms and using the fact that the norm of 2 is 4, we obtain

$$4 = (a^2 + 5b^2)(c^2 + 5d^2).$$

If neither of the elements $a + b\sqrt{-5}$ and $c + d\sqrt{-5}$ is a unit, their norms must be greater than 1, and the preceding equation shows that each must have norm 2. But it is readily verified that there do not exist integers a and b such that $a^2 + 5b^2 = 2$; hence no element of $\mathbf{Z}[\sqrt{-5}]$ has norm 2. We conclude that 2 is an irreducible in $\mathbf{Z}[\sqrt{-5}]$. Similarly, it may be verified that there is no element of norm 3, and therefore 3 is also an irreducible element by the same kind of argument. Since 2 and 3 are irreducible in $\mathbf{Z}$, we might be tempted to guess that every irreducible in $\mathbf{Z}$ is also irreducible when considered as an element of $\mathbf{Z}[\sqrt{-5}]$. However, this would be false since $41 = (6 + \sqrt{-5}) \times (6 - \sqrt{-5})$.

One of the most interesting features of the integral domain $\mathbf{Z}[\sqrt{-5}]$ is that there exist in this domain elements which have different factorizations into a product of irreducibles—actually different in the sense that the irreducibles in different factorizations are not necessarily associates. As an example, consider the following two factorizations of 6 in the domain $\mathbf{Z}[\sqrt{-5}]$:

$$6 = 2 \cdot 3 \quad \text{and} \quad 6 = (1 + \sqrt{-5})(1 - \sqrt{-5}).$$

We proved above that 2 is an irreducible and indicated that a similar argument would show that 3 is irreducible. Now $N(1 \pm \sqrt{-5}) = 6$ and if $1 \pm \sqrt{-5}$ were not irreducible, there would have to exist an element of norm 2 and an element of norm 3. Since there are no such elements, we conclude that $1 \pm \sqrt{-5}$ is irreducible. Moreover, since the only units of $\mathbf{Z}[\sqrt{-5}]$ are ± 1, it is clear that no one of the elements $2, 3, 1 + \sqrt{-5}, 1 - \sqrt{-5}$ is an associate of another; hence we do indeed have different factorizations of 6 into a product of irreducibles. In particular, property (ii) of Definition 11.9 does not hold, and $\mathbf{Z}[\sqrt{-5}]$ is therefore not a U.F.D.

We may point out that the above calculations show that no one of the irreducible elements $2, 3, 1 + \sqrt{-5}, 1 - \sqrt{-5}$ is a prime. For example, 2 clearly divides $(1 + \sqrt{-5})(1 - \sqrt{-5}) = 6$, but does not divide either of these factors. A similar observation shows that all of the other mentioned irreducibles are not primes.

11.2 PRINCIPAL IDEAL DOMAINS

We recall (Definition 10.2) that in a commutative ring with unity, an ideal generated by a single element is called a *principal ideal.* If $a \in D$, where D is an integral domain, we shall now find it convenient to denote the principal ideal generated by an element a by aD, that is,

$$aD = \{ax \mid x \in D\},$$

and aD consists simply of all multiples of a in D.

We proved in Theorem 10.3 that every ideal in $\mathbf{Z}$ is a principal ideal, and that the same is true for a polynomial ring $F[x]$ over a field F. The following terminology will be convenient in referring to such domains.

11.11 Definition. An integral domain D is called a *principal ideal domain* if every ideal in D is a principal ideal.

Thus we can say that **Z** and $F[x]$ are principal ideal domains. The rest of this chapter will be devoted to sketching a proof of the following significant result.

11.12 Theorem. *A principal ideal domain is a unique factorization domain.*

In view of this result, the unique factorization theorems in **Z** and in $F[x]$ may both be considered to be consequences of the fact that they are principal ideal domains.

It will be pointed out in the notes at the end of the chapter that the converse of Theorem 11.12 is not true, that is, that there exist unique factorization domains which are not principal ideal domains.

Throughout the remainder of this chapter D will be a principal ideal domain with unity e. We shall establish several lemmas from which the proof of Theorem 11.12 will follow fairly easily.

First, let us point out that the following result is an immediate consequence of Lemma 11.5 (whether or not D is a principal ideal domain).

11.13 Lemma. *If a and b are nonzero elements of the integral domain D, then $aD = bD$ if and only if a and b are associates. Moreover, $aD = eD = D$ if and only if a is a unit.*

The next lemma will play an important role in the proof of our theorem, although at first glance it may not appear to have any connection with it.

11.14 Lemma. *In a principal ideal domain D there cannot exist an infinite sequence of ideals a_1D, a_2D, a_3D, $\cdots$ such that each is properly contained in the following; that is, we cannot have*

11.15
$$a_1D \subset a_2D \subset a_3D \subset \cdots.$$

PROOF. Suppose, on the contrary, that there does exist an infinite sequence of ideals satisfying 11.15, and let us seek a contradiction.

First, we observe that the union A of all the ideals in 11.15 is itself an ideal in D. For if $a, b \in A$, we have that $a \in a_iD$ and $b \in a_jD$ for certain positive integers i and j. For convenience of notation, let us assume that $i \leq j$. Then $a_iD \subset a_jD$ and hence, in particular, both a and b are in the ideal a_jD. Thus $a + b \in a_jD \subseteq A$.

Moreover, if $a \in A$, say $a \in a_iD$, and $c \in D$, then $-a \in a_iD = A$ and $ac \in a_iD \subseteq A$. This shows that A is an ideal in D.

Since D is a principal ideal domain, the ideal A which is the union of the ideals in 11.15 is a principal ideal, say, $A = xD$ for some $x \in D$. But $x \in A$ and therefore x is in some one of the ideals whose union is A, that is, $x \in a_kD$ for some positive integer k. Hence $A = xD \subseteq a_kD$ and from this it follows that $a_jD \subseteq A \subseteq a_kD$ for every integer $j \geq k$. This gives the desired contradiction and completes the proof.

If it is true in a given ring that there cannot exist an infinite sequence of ideals such that each is properly contained in the following one, we often express this fact by saying that the *ascending chain condition* holds. The lemma just proved therefore shows that the ascending chain condition holds in every principal ideal domain.

Another concept which we shall find useful is the following.

11.16 Definition. An ideal A in an arbitrary ring R is said to be a *maximal ideal* if $A \neq R$ and there exists no ideal B such that $A \subset B \subset R$.

We can now prove the next lemma, which indicates the close connection between the concept of maximal ideal and that of an irreducible element of a principal ideal domain.

11.17 Lemma. *A nonzero ideal aD in a principal ideal domain D is a maximal ideal if and only if a is an irreducible of D.*

PROOF. Suppose, first, that a is irreducible and let us assume that aD is not a maximal ideal and seek a contradiction. By our assumption, there exists an ideal bD of D such that $aD \subset bD \subset D$. The fact that $aD \subset bD$ implies that $b|a$ and that b is not an associate of a (by Lemma 11.13). Moreover, $bD \subset D$ implies that b is not a unit. We have therefore shown that $b|a$ and that b is neither a unit nor an associate of a. This shows that a is not irreducible, and we have the desired contradiction. We have therefore shown that if a is irreducible, then aD is a maximal ideal in D.

Conversely, suppose that aD is a maximal ideal in D. If a were not irreducible, it would have a divisor d, neither a unit nor an associate of a. But this would imply that $aD \subset dD \subset D$, violating our hypothesis that aD is a maximal ideal. We conclude that a must be irreducible and the proof is complete.

We are now ready to prove the following step in the proof of our factorization theorem.

11.18 Lemma. *If the nonzero element a of a principal ideal domain D is not a unit, it has an irreducible divisor.*

PROOF. If a itself is irreducible, there is nothing to prove. If a is not irreducible, the preceding lemma shows that aD is not a maximal ideal. Thus there exists an ideal a_1D such that $aD \subset a_1D \subset D$. It follows that $a_1|a$ and that a_1 is not a unit. If a_1 is not irreducible, we can apply the same argument to a_1 and conclude that there exists a nonunit a_2 which divides a_1 and is not an associate of a_1. Accordingly, we would have that $aD \subset a_1D \subset a_2D \subset D$. Similarly, if a_2 is not irreducible, it has a nonunit divisor a_3 which is not an associate of a_2, and therefore $aD \subset a_1D \subset a_2D \subset a_3D \subset D$. By Lemma 11.14, we cannot have an infinite sequence of this kind; therefore the above procedure must come to an end after a finite number of steps. Thus, there must exist an irreducible divisor a_n of a_{n-1} for some positive integer n and it is clearly also a divisor of a. This completes the proof.

Using this lemma and an argument somewhat like that just used in the proof of the lemma, let us now prove the first property required in showing that a principal ideal domain is a unique factorization domain; that is, we shall prove the following result.

11.19 Lemma. *If the nonzero element a of a principal ideal domain D is not a unit, there exist irreducibles $p_1, p_2, \cdots, p_r$ of D (r a positive integer) such that*

11.20 $$a = p_1p_2 \cdots p_r.$$

PROOF. Since a is not a unit, the preceding lemma shows that a has an irreducible divisor p_1, and $a = p_1c_1$ for some nonzero element c_1 of D. We therefore have $aD \subset c_1D$. Now if c_1 is not a unit, by applying the same argument to c_1 as we have just applied to a, there exists an irreducible divisor p_2 of c_1. If we write $c_1 = p_2c_2$, then $a = p_1p_2c_2$ and if c_2 is not a unit, we have $aD \subset c_1D \subset c_2D$, and c_2 has an irreducible divisor p_3. At this stage, we would have $a = p_1p_2p_3c_3$ for some element c_3. This same procedure can be applied as long as c_i is not a unit. Lemma 11.14 asserts that there must exist a positive integer r such that c_r is a unit since, otherwise, we would have an infinite sequence of ideals

$$aD \subset c_1D \subset c_2D \subset \cdots.$$

Thus we have

$$a = p_1p_2p_3 \cdots p_rc_r,$$

where the p's are irreducibles and c_r is a unit. But then p_rc_r is an irreducible and we might just as well change the notation and call it p_r. This gives the desired result 11.20.

We have thus established the first property required in proving that a principal ideal domain is a unique factorization domain. We shall now briefly present some results from which the second property (that of uniqueness) can be obtained by a simple adaptation of the proof of Theorem 4.20 for the integers. The details of the proof will then be left as an exercise.

If a and b are nonzero elements of a principal ideal domain D, an element d of D is said to be a *greatest common divisor* (g.c.d) of a and b if the following two conditions hold:

(i) $d|a$ and $d|b$.
(ii) If $c \in D$ such that $c|a$ and $c|b$, then $c|d$.

It will be observed that if d is a g.c.d. of a and b, so also is any associate of d. In the case of the integers, we found it convenient to define a *unique* g.c.d. by requiring that it be positive. Similarly, for polynomials over a field we obtained a *unique* g.c.d. by requiring that it be a monic polynomial. However, for an arbitrary principal ideal domain there is no obvious way to get a unique g.c.d., so we simply get along without uniqueness since it is not essential anyway. The existence of a g.c.d. of two nonzero elements of a principal ideal domain will presently be established.

If $a, b \in D$, it is easy to verify that the set

11.21 $$\{ax + by \mid x, y \in D\}$$

is an ideal in D, and is therefore a principal ideal since in D every ideal is principal. Let us now prove the following result.

11.22 Lemma. *If a and b are nonzero elements of the principal ideal domain D, any generator d of the ideal 11.21 is a g.c.d. of a and b.*

PROOF. Since

$$dD = \{ax + by \mid x, y \in D\},$$

we see that $a \in dD$ and $b \in dD$. That is, $d|a$ and $d|b$; and thus d satisfies the first defining condition for a g.c.d. of a and b. The other condition also follows quite easily, for since $d \in D$, there exist elements x_1 and y_1 of D such that $d = ax_1 + by_1$. From this equation, we see at once that if $c|a$ and $c|b$, then $c|d$. Thus both conditions are satisfied and the proof is complete. In particular, this lemma shows the existence of a g.c.d. for any two nonzero elements of a principal ideal domain.

The next result furnishes an essential step in proving the uniqueness of factorization which we still need to prove to show that D is a U.F.D.

11.23 Lemma. *In a principal ideal domain an irreducible element is a prime element.*

PROOF. Suppose that c is an irreducible such that $c|(ab)$ and $c \nmid a$, and let us prove that $c|b$. Since $c \nmid a$, a g.c.d. of c and a cannot be an associate of the irreducible c and therefore must be a unit. But the ideal generated by a unit of D is D itself. In view of the preceding lemma, we see that

$$D = \{ax + cy \mid x, y \in D\}$$

and, in particular, if e is the unity of D, there exist elements x_1 and y_1 of D such that $ax_1 + cy_1 = e$. Multiplying by b we obtain

$$(ab)x_1 + c(by_1) = b.$$

Since it is given that $c|(ab)$, it is clear from this equation that c divides the left side of this equation, and we conclude that $c|b$. This shows that c is prime and completes the proof of the lemma.

In view of this lemma, we see that the concept of irreducible element and prime element are equivalent in any principal ideal domain, as we already knew for **Z** and for $F[x]$, where F is a field.

It will be observed that, after developing the appropriate machinery, the proof just given follows very closely the proof of Lemma 4.18 for the integers.

EXERCISES

1. In $\mathbf{Z}[\sqrt{-5}]$, verify that $21 = 3 \cdot 7 = (1 + 2\sqrt{-5})(1 - 2\sqrt{-5})$, and that these are two different factorizations of 21 into a product of irreducibles.
2. Prove by induction that if an irreducible in a principal ideal domain divides a product of any finite number n of elements of D, it must divide at least one of these elements.
3. Use the result of the preceding exercise and the method of proof of Theorem 4.20 to complete the proof (Theorem 11.12) that a principal ideal domain is a unique factorization domain.
4. Define a least common multiple (l.c.m.) of two nonzero elements a and b of a principal ideal domain. Prove that if $aD \cap bD = mD$, then m is a l.c.m. of a and b.

NOTES AND REFERENCES

For a detailed discussion of a number of integral domains similar to the domain $\mathbf{Z}[\sqrt{-5}]$ (the so-called quadratic domains), see Hardy and Wright [19], Chapters 14 and 15.

It is known that if D is an U.F.D., then so also is the polynomial domain $D[x]$. For a proof of this fact, see Fraleigh [5], p. 266. It was shown near the end of Section 10.1 that $\mathbf{Z}[x]$ is not a principal ideal domain. Since $\mathbf{Z}$ is a U.F.D., $\mathbf{Z}[x]$ is also a U.F.D. which is not a principal ideal domain.

For an example of an integral domain in which there are elements which are not units and which cannot be expressed in any way as a product of irreducibles, look at Exercise 2, p. 118 of Jacobson, *Lectures in Abstract Algebra*, vol. 1, Van Nostrand, New York, 1951.

Bibliography

Many of the books listed below contain bibliographies which may be used to supplement this somewhat limited one. We have listed here a small sample of books at least parts of which should be quite readable by students of this text. Most of them include a number of topics not covered in this book. In particular, except for minor modifications, item [8] is an extension of this book and the approach to the subject coincides with that of this text. The other books on abstract algebra in general should give the interested student a variety of approaches to the subject. Particularly recommended for those with a minimal mathematical background are [4], [11], and [12]. Comments about the more specialized books listed below will be found in notes at the ends of appropriate chapters.

Abstract Algebra in General

1. Ames, Dennis B., *An Introduction to Abstract Algebra*, International, Scranton, Pa., 1969.
2. Birkhoff, Garrett, and Saunders MacLane, *A Survey of Modern Algebra* (3rd ed.), Macmillan, New York, 1965.
3. Dean, Richard A., *Elements of Abstract Algebra*, Wiley, New York, 1966.
4. Dubisch, Roy, *Introduction to Abstract Algebra*, Wiley, New York, 1965.
5. Fraleigh, John B., *A First Course in Abstract Algebra*, Addison-Wesley, Reading, Mass., 1967.
6. Herstein, I. N., *Topics in Algebra*, Blaisdell, New York, 1964.
7. Johnson, Richard E., *University Algebra*, Prentice-Hall, Englewood Cliffs, N.J., 1966.
8. McCoy, Neal H., *Fundamentals of Abstract Algebra*, Allyn and Bacon, Boston, 1972.

9. Mostow, George D., Joseph H. Sampson, and Jean-Pierre Meyer, *Fundamental Structures of Algebra*, McGraw-Hill, New York, 1963.
10. Paley, Hiram, and Paul M. Weichsel, *A First Course in Abstract Algebra*, Holt, Rinehart and Winston, New York, 1966.
11. Perlis, Sam, *Introduction to Algebra*, Blaisdell, Waltham, Mass., 1966.
12. Whitesitt, J. E., *Principles of Modern Algebra* (2nd ed.) Addison-Wesley, Reading, Mass., 1973.
13. Zariski, Oscar, and Pierre Samuel, *Commutative Algebra*, vol. 1, Van Nostrand, Princeton, N.J., 1958.

Group Theory

14. Baumslag, Benjamin, and Bruce Chandler, *Theory and Problems of Group Theory*, Schaum's Outline Series, McGraw-Hill, New York, 1968.
15. Hall, Marshall, Jr., *The Theory of Groups* (rev. ed.) Macmillan, New York, 1961.
16. Ledermann, Walter, *Introduction to the Theory of Finite Groups* (4th ed.), Interscience, New York, 1961.
17. Rotman, Joseph H., *The Theory of Groups: An Introduction* (2nd ed.), Allyn and Bacon, Boston, 1973.

Number Theory

18. Davenport, H., *The Higher Arithmetic* (3rd ed.), Hutchinson's University Library, London, 1968.
19. Hardy, G. H., and E. M. Wright, *An Introduction to the Theory of Numbers* (4th ed.), Oxford University Press, Oxford, 1960.
20. LeVeque, William J., *Elementary Theory of Numbers*, Addison-Wesley, Reading, Mass., 1962.
21. Long, Calvin T., *Elementary Introduction to Number Theory* (2nd ed.), Heath, Boston, 1972.
22. Ore, Oystein, *Number Theory and its History*, McGraw-Hill, New York, 1948.
23. Sierpinski, W., *Elementary Theory of Numbers*, translated by A. Hulanicki, Hafner, New York, 1964.

Ring Theory

24. Burton, David M., *A First Course in Rings and Ideals*, Addison-Wesley, Reading, Mass., 1970.
25. Herstein, I. N., *Noncommutative Rings* (Carus Mathematical Monograph no. 15), The Mathematical Association of America, Washington, D.C., 1968.
26. McCoy, Neal H., *Rings and Ideals* (Carus Mathematical Monograph no. 8), The Mathematical Association of America, Washington, D.C., 1948.
27. McCoy, Neal H., *The Theory of Rings*, Chelsea, New York, 1973.

Index

D

E

F

G

N

O

P

Q

R

S